1990

Numerical Methods for Initial Value Problems
in Ordinary Differential Equations

This is a volume in
COMPUTER SCIENCE AND SCIENTIFIC COMPUTING

Werner Rheinboldt and Daniel Siewiorek, editors

Numerical Methods for Initial Value Problems in Ordinary Differential Equations

Simeon Ola Fatunla
Department of Mathematics
University of Benin
Benin City
Nigeria

ACADEMIC PRESS, INC.
Harcourt Brace Jovanovich, Publishers

Boston   San Diego   New York
Berkeley   London   Sydney   Tokyo   Toronto

ACADEMIC PRESS, INC.
1250 Sixth Avenue, San Diego, CA 92101

United Kingdom Edition published by
ACADEMIC PRESS INC. (LONDON) LTD.
24-28 Oval Road, London NW1 7DX

Library of Congress Cataloging-in-Publication Data

Fatunla, Simeon Ola.
  Numerical methods for initial value problems in
ordinary differential equations.

  (Computer science and scientific computing)
  Bibliography: p.
  Includes index.
  1. Differential equations—Numerical solutions.
2. Initial value problems—Numerical solutions.
I. Title. II. Series.
QA372.F35 1987   515.3'5   87-18695
ISBN 0-12-249930-1

88 89 90 91 9 8 7 6 5 4 3 2 1

Printed in the United States of America

CONTENTS

PREFACE   ix

1   PRELIMINARIES   1

    1.1 The Difference Operators   1
    1.2 Theory of Interpolation   3
    1.3 Finite Difference Equations   10
    1.4 Linear Systems with Constant Coefficients   13
    1.5 Distribution of Roots of Polynomials   15
    1.6 First Integral Mean Value Theorem   18
    1.7 Common Norms in ODEs   18

2   NUMERICAL INTEGRATION ALGORITHMS   21

    2.1 Introduction   21
    2.2 Existence of Solution, Numerical Approach   25
    2.3 Special IVPs   27
    2.4 Error Propagation, Stability and
         Convergence of Discretization Methods   30

3   THEORY OF ONE-STEP METHODS   33

    3.1 General Theory of One-Step Methods   33
    3.2 The Euler Scheme, the Inverse Euler Scheme
         and Richardson's Extrapolation   37
    3.3 The Convergence of Euler's Scheme   41
    3.4 The Trapezoidal Scheme   44

4   RUNGE-KUTTA PROCESSES   51

    4.1 General Theory of Runge-Kutta Processes   51
    4.2 The Explicit Two-Stage Process   59
    4.3 Convergence and Stability of Two-Stage
         Explicit R-K Scheme   64
    4.4 Matrix Representation of the R-K Processes   66
    4.5 Error Estimation and Stepsize Selection

in R-K Processes   76
4.6 Implicit and Semi-Implicit R-K Processes   79
4.7 Rosenbrock Methods   89

5   LINEAR MULTISTEP METHODS   93

5.1 Starting Procedure   93
5.2 Explicit Linear Multistep Methods   95
5.3 Implicit Linear Multistep Methods   100
5.4 Implementation of the Predictor-Corrector Formulas   105
5.5 General Theory of Linear Multistep Methods   109
5.6 Automatic Implementation of the Adams Scheme   127

6   NUMERICAL TREATMENT OF SINGULAR/DISCONTINUOUS
            INITIAL VALUE PROBLEMS   131

6.1 Introduction   131
6.2 Non-Polynomial Methods   133
6.3 The Inverse Polynomial Methods   140
6.4 Local Error Estimates in Automatic
       Codes for Discontinuous Systems   145

7   EXTRAPOLATION PROCESSES AND SINGULARITIES   147

7.1 Introduction   147
7.2 Generation of the Zero-th Column of
       Extrapolation Table   149
7.3 Polynomial and Rational Extrapolation   155
7.4 Convergence and Stability Properties
       of Extrapolation Processes   163
7.5 Practical Implementation of
       Extrapolation Processes   167

8   STIFF INITIAL VALUE PROBLEMS   169

8.1 The Concept of Stiffness   169
8.2 Stiff and Nonstiff Algorithms   175

8.3 Solution of Nonlinear Equations

and Estimation of Jacobians   177
8.4 Region of Absolute Stability   181
8.5 Stability Criteria for Stiff Methods   189
8.6 Stronger Stability Properties of IRK Processes   198
8.7 One-Leg Multistep Methods   206

9   STIFF ALGORITHMS   209

9.1 What are Stiff Algorithms   209
9.2 Efficient Implementation of Implicit
     Runge-Kutta Methods   214
9.3 The Backward Differentiation Formula   223
9.4 Second Derivative Formulas   228
9.5 Extrapolation Processes for Stiff Systems   230
9.6 Mono-Implicit Runge-Kutta Methods   236

10   SECOND ORDER DIFFERENTIAL EQUATIONS   239

10.1 Introduction   239
10.2 Linear Multistep Methods and the
      Concept of P-Stability   241
10.3 Derivation of P-Stable Formulas   244
10.4 One-Leg Multistep Methods for Second Order IVPs   245
10.5 Multiderivative Methods   250

11   RECENT DEVELOPMENTS IN ODE SOLVERS   253

REFERENCES

PREFACE

The mathematical models of physical phenomena in many spheres of human endeavor often yield systems of ordinary differential equations. The finite difference method has excelled for the numerical treatment of such problems since the advent of digital computers about four decades ago. The development of algorithms has been largely guided by convergence and stability theorems of Dahlquist (1956, 1963, 1978b) as well as the treatises of Henrici (1962) and Stetter (1973).

This monograph is essentially meant to complement the earlier works of Gear (1971b) and Lambert (1973a). Advanced undergraduates in mathematics, computer science, and engineering should derive sufficient benefits from the early chapters of this book. Postgraduate students and researchers in related disciplines may find the book invaluable.

Although the monograph is, to a large extent, exhaustive, readers who have done an introductory course in numerical analysis, linear algebra, and calculus should find the material very beneficial. An extensive bibliography is provided to guide the prospective readers.

Emphasis has been placed on the numerical treatment of special differential equations: stiff, stiff oscillatory, singular, and discontinuous initial value problems, which are often characterized by large Lipschitz constants and which have received considerable attention in the last two decades. Existing robust codes for these categories of problems are highlighted.

The early chapters of this book (Chapters 1-5) contain materials for a one-semester course given to final year undergraduate students in mathematics at the Trinity College, Dublin, Ireland, and the University of Benin, Benin City, Nigeria. Chapters 6-11 were given in the form of seminars to postgraduate students at both institutions.

Chapter 1 identifies the basic mathematical tools which will enable the reader to follow the contents of the monograph. Chapter 2 discusses the existence and uniqueness theorem and the propagation of errors. Chapter 3 treats the general theory of one-step methods. Chapter 4 dwells on the Runge-Kutta processes with particular mention of Rosenbrock schemes and Block Runge-Kutta methods. Chapter 5 deals with the development and theory of the linear multistep methods. Mentioned in the last section of Chapter 5 are robust codes based on Adams-Moulton methods.

In Chapter 6 we examine some recent developments in the numerical treatment of singular, and discontinuous initial value problems. In Chapter 7 we consider the application of extrapolation processes to both nonstiff, stiff, and singular initial value problems. The semi-implicit midpoint rule combined with polynomial extrapolation (Bader and Deuflhard 1983) is treated in this chapter and Chapter 9.

Chapter 8 discusses the concept of stiffness and various stability properties, including the stronger stability criteria peculiar to implicit R-K processes: B-stability, BN-stability, AN-stability, algebraic stability, and the solution of nonlinear equations, which are always associated with implicit methods.

Chapter 9 identifies the properties of stiff algorithms, practical (automatic) implementation of the Runge-Kutta schemes, backward differentiation formulas, second derivative methods, the extrapolation processes, and cyclic multistep methods. Currently available robust codes are identified.

Chapter 10 is concerned with the numerical solution of second order initial value problems. The symmetric multistep methods and P-stable schemes—multistep, multiderivative methods, one-leg multistep, and hybrid methods—are discussed.

Chapter 11 examines some recent developments in ODE solvers and highlights potential research areas.

Many people contributed towards the preparation of this book. It is impossible to mention all, and so I can only thank a few here. The others will bear with me for lack of space!

I am indeed grateful to Professor Robert D. Skeel, who painstakingly read the initial draft and offered invaluable suggestions that helped to shape the contents. I am particularly indebted to Professor C. William Gear and Dr. Sandy Gourlay, whose suggestions on a few of the chapters improved their readability. The latter also introduced me to ordinary differential equations.

My special thanks to Professor J.J.H. Miller, of the School of Mathematics, Trinity College, Dublin, Ireland, for accommodating me in his department during my sabbatical leave (1981-1982), during which period the first half of the book was completed. I am equally grateful to Professors C. W. Gear and R. D. Skeel, both of the Department of Computer Science, University of Illinois at

Urbana-Champaign, for accommodating me in their faculty in the summers of 1984, 1986, and 1987. They were indeed short but eventful and memorable periods.

Special thanks also go to Professors Lawrence F. Shampine, Hans J. Stetter, M. Van Veldhuizen, Germund Dahlquist, J.R. Cash, Wayne Enright, Bengt Lindberg, Guyne Evans, and Arieh Iserles, whose comments on the final draft greatly improved the readability of the entire text.

I thank my students who served as guinea pigs: Conor Fitzsimons, Nial Ferguson, Daniel Okunbor, and Raphael Okojie. I express my appreciation to Barbara Armstrong for her excellent typing from an almost unintelligible script, and to Hon-Wah Tam for his technical assistance during the final preparation of the monograph.

I am really indebted to my inspiring vice-chancellor, Professor Grace Alele Williams, for her support and encouragement; my close confidant, friend and colleague, Professor S. O. Iyahen; my close friends and guardians, Jean and Dick Toward; and my parents, Janet and Joshua Fatunla.

Finally, I thank my wife, Grace, for her patience and support during my one year sabbatical leave at Trinity College, Dublin, Ireland, when the bulk of the material was prepared.

Simeon Ola Fatunla

1. PRELIMINARIES

1.1. The Difference Operators

Let $f(x)$ be a function defined on an interval [a, b] of the real line and $f_{n+j} = f(x_n + j \cdot h)$ for specified real numbers x_n, h, and integer j, h being the mesh interval.

The shift operator E is defined as

$$E^j f_n = f(x_n + j \cdot h), \; j \text{ integer} . \qquad (1.1.1)$$

Also, the backward difference operator ∇ is defined as

$$\nabla f_n = f_n - f_{n-1} = (1 - E^{-1}) f_n ,$$

thus leading to the relation

$$\nabla = (1 - E^{-1}) \qquad (1.1.2)$$

between the operators E and ∇.

The Relation (1.1.2) is quite useful in the derivation of higher order backward differences such as

$$\nabla^j f_n = (1 - E^{-1})^j f_n$$

$$= \sum_{r=0}^{j} (-1)^r \binom{j}{r} E^{-r} f_n \qquad (1.1.3)$$

1

$$= \sum_{r=0}^{j} (-1)^r \binom{j}{r} f_{n-r} \, ,$$

where

$$\binom{j}{r} = \begin{cases} 1, & r = 0 \\ \\ \dfrac{j!}{r!(j-r)!} \, , & r \neq 0. \end{cases}$$

In an identical manner, the forward difference operator Δ is given by

$$\Delta = E - 1 \, , \qquad\qquad (1.1.4)$$

and the higher forward difference operator Δ^j operating on f_n gives

$$\Delta^j f_n = \sum_{r=0}^{j} (-1)^r \binom{j}{r} f_{n+r} \, . \qquad\qquad (1.1.5)$$

The shift operator can admit non-integral powers j such that

$$E^j f_n = E^j f(x_n) = f(x_n + j \cdot h) \, , \ j \text{ real} \, . \qquad\qquad (1.1.1')$$

Finally, the central difference operator δ is defined as

$$\delta f_n = f_{n+1/2} - f_{n-1/2} \, .$$

This implies

$$\delta = E^{1/2} - E^{-1/2} \, , \qquad\qquad (1.1.6)$$

and, in general, we have

$$\delta^{2j} f_n = \sum_{r=0}^{2j} (-1)^r \binom{2j}{r} f_{n+j-r} \, . \qquad\qquad (1.1.7)$$

For example,

$$\delta^2 f_n = f_{n+1} - 2f_n + f_{n-1} \, ,$$

and

$$\delta^4 f_n = f_{n+2} - 4f_{n+1} + 6f_n - 4f_{n-1} + f_{n-2} \, .$$

We now state an important theorem of difference calculus.

Theorem 1.1. *If $f(x)$ is a polynomial of degree $\leq k$, then*

$$\nabla^{k+1} f(x) = 0 ,$$

$$\Delta^{k+1} f(x) = 0 , \qquad (1.1.8)$$

$$\delta^{k+1} f(x) = 0 .$$

Proof. The proof follows from the definition of the difference operators, coupled with the fact that these operators are distributive over addition in the sense that

$$\nabla^j [\alpha f(x) + \beta g(x)] = \alpha \nabla^j f(x) + \beta \nabla^j g(x) . \qquad (1.1.9)$$

1.2. Theory of Interpolation

The theory of interpolation was of immense importance in the pre-computer era, when the cost of function evaluation could be enormous. Even in this computer age, interpolation theory is still of importance in many areas of numerical analysis, including the development of linear multistep methods for initial value problems in ordinary differential equations.

Given a function $f(x)$ whose graph passes through the set of distinct points $\{(x_i, f(x_i)) \mid i = 0(1)k\}$, the function $f(x)$ can be approximated by a polynomial $P_k(x)$ of degree $\leq k$ such that

$$P_k(x_i) = f(x_i) \quad i = 0(1)k .$$

Three different ways of writing a polynomial $P_k(x)$ of degree k are

(i)   Lagrangian interpolation,

(ii) Newton or divided difference interpolation,

(iii) a linear combination of monomial bases.

The Lagrange interpolation formula $P_k(x)$ of degree $\leq k$ is specified as

$$P_k(x) = \sum_{i=0}^{k} l_i(x) f_i , \quad f_i = f(x_i) , \qquad (1.2.1)$$

where each

$$l_i(x) = \prod_{\substack{j=0 \\ j \neq i}}^{k} \left(\frac{x - x_j}{x_i - x_j} \right) \qquad (1.2.2)$$

is a polynomial of degree k with

$$l_i(x_j) = \delta_{ij} , \qquad (1.2.3)$$

where δ_{ij} is the well known Kronecker delta defined as

$$\delta_{ij} = \begin{cases} 0 & i \neq j \\ \\ 1 & i = j \end{cases} . \qquad (1.2.4)$$

An error estimate for the Lagrangian interpolation polynomial is given by the following theorem:

Theorem 1.2. *Let $\{x_i, i = 0(1)k\}$ be distinct points on the real line and $f_i = f(x_i)$ $i = 0(1)k$ with $f(x_i)$ a real valued function, sufficiently differentiable. There exists a point ξ between the least and greatest of $x_0, x_1, ..., x_k, t$ such that the error $E(t) = f(t) - P_k(t)$ in (1.2.1) is given by*

$$E(t) = \frac{\omega_k(t)}{(k+1)!} f^{(k+1)}(\xi) , \qquad (1.2.5)$$

$\omega_k(x)$ *being a polynomial of degree $k+1$ given as*

$$\omega_k(x) = \prod_{i=0}^{k} (x - x_i) . \qquad (1.2.6)$$

Proof. If $t = x_i$, $i = 0(1)k$, $\omega_k(t) = 0$, and, hence, (1.2.5) is satisfied. Now assume $t \neq x_i$ for $i = 0(1)k$ and define $G(x)$ as

$$G(x) = E(x) - \frac{\omega_k(x)}{\omega_k(t)} E(t) . \qquad (1.2.7)$$

$G(x)$, $E(x)$, and $\omega_k(x)$ all have $k+1$ continuous derivatives. We insert the identities

$$E^{(k+1)}(x) = f^{(k+1)}(x) , \quad \omega_k^{(k+1)}(x) = (k+1)! \qquad (1.2.8)$$

in $G^{(k+1)}(x)$ to get

$$G^{(k+1)}(x) = f^{(k+1)}(x) - \frac{(k+1)!}{\omega_k(t)} E(t) . \qquad (1.2.9)$$

$G(x)$ has $k+2$ zeros at $\{x_0, x_1, ..., x_k, t\}$, and, hence, $k+1$ applications of Rolle's theorem (Philips 1962) imply the existence of a point $\xi \in [x_0, x_1, ..., x_k, t]$ such that $G^{(k+1)}(\xi) = 0$. This, in (1.2.9), implies (1.2.5). ∎

Exercise. Show that

$$l_k(x) = \frac{\omega_k(x)}{(x-x_k)\omega_k'(x_k)} .$$

The main snag with the Lagrangian interpolation scheme is that it does not accommodate an easy change in the degree of the polynomial. The Newton divided difference scheme takes care of this deficiency and can be generated recursively as follows:

Set

$$Q_0(x) = f_0$$

and define

$$Q_1(x) = Q_0(x) + (x-x_0)a_1 ,$$

with the constraints

$$Q_1(x_0) = f_0 , \quad Q_1(x_1) = f_1 ,$$

yielding

$$a_1 = \frac{f_1 - f_0}{x_1 - x_0} .$$

We denote the quantity $f_1 - f_0 / x_1 - x_0$ by $f[x_0, x_1]$.

In general, for $k \geq 0$,

$$Q_k(x) = Q_{k-1}(x) + \prod_{j=0}^{k-1} (x-x_j)a_k , \tag{1.2.10}$$

with the requirements that

$$Q_k(x_j) = f_j , \quad j = 0(1)k ,$$

thus yielding

$$a_k = \frac{f_k - Q_{k-1}(x_k)}{\prod_{j=0}^{k-1} (x_k - x_j)} \tag{1.2.11}$$

$$= f[x_0, x_1, x_2, ..., x_k] .$$

Exercise. Show that

$$f[x_0, x_1, ..., x_k] = \frac{f[x_0, ..., x_{k-1}] - f[x_1, ..., x_k]}{x_0 - x_k} .$$

The Newton or divided difference interpolation polynomial of degree k is hence given as

$$Q_k(x) = f_0 + (x-x_0)f[x_0, x_1] + (x-x_0)(x-x_1)f[x_0, x_1, x_2]$$

$$(1.2.12)$$

$$+ ... + \prod_{j=0}^{k-1} (x-x_j)f[x_0, x_1, ..., x_k] .$$

Lemma 1.1. *There exists a unique polynomial of degree k with prescribed values at $k+1$ points (x_i, f_i), $i = 0(1)k$ if $x_i \neq x_j$, $i \neq j$.*

Proof. Let

$$R(x) = P_k(x) - Q_k(x) \qquad (1.2.13)$$

denote a polynomial of degree at most k. From (1.2.1), (1.2.2), and (1.2.12), we have

$$R(x_i) = 0 , \quad i = 0(1)k ,$$

thus giving a polynomial of degree $\leq k$ with $k+1$ zeros at distinct points x_i, $i = 0(1)k$.

Hence

$$R(x) \equiv 0 .$$

∎

We state the following theorem:

Theorem 1.3. *The Lagrange interpolation scheme and the divided difference scheme are equivalent.*

Proof. Both $P_k(x)$ given by (1.2.1) and $Q_k(x)$ specified by (1.2.12) are polynomials of degree k satisfying

$$P_k(x_i) = Q_k(x_i) = f_i , \quad i = 0(1)k . \qquad (1.2.14)$$

The result follows from Lemma 1.1.

∎

Corollary 1.1.

$$f[x_0, x_1, ..., x_k] = \sum_{i=0}^{k} \left(\prod_{\substack{j=0 \\ j \neq i}}^{k} \frac{1}{(x_i - x_j)} \right) f_i . \qquad (1.2.15)$$

Proof. The Lagrange interpolating polynomial $P_k(x)$ specified by (1.2.1) and (1.2.2) can be rearranged as

$$P_k(x) = \sum_{i=0}^{k} (\prod_{\substack{j=0 \\ j \neq i}}^{k} (\frac{x-x_j}{x_i-x_j}))f_i$$

(1.2.16)

$$= x^k \sum_{i=0}^{k} (\prod_{\substack{j=0 \\ j \neq i}}^{k} \frac{1}{(x_i-x_j)})f_i + \text{lower degree terms} .$$

Similarly, the divided difference formula (1.2.12) can be expressed as

$$Q_k(x) = f_0 + (x-x_0)f[x_0, x_1]$$

$$+ ... + (x-x_0)(x-x_1)...(x-x_{k-1})f[x_0, x_1, ..., x_k]$$

(1.2.17)

$$= x^k f[x_0, x_1, ..., x_k] + \text{lower degree terms} .$$

Equating coefficients of x^k in Equations (1.2.16) and (1.2.17) implies Equation (1.2.15). ∎

The Newton or divided difference interpolation is often preferred to the Lagrange interpolation because it admits lowering or raising the degree of the interpolating polynomial with minimal computational effort. However, each representation has advantages which may make it more suitable than others in a given application.

It is, at times, more convenient to generate the interpolating polynomial in the reversed form or to rearrange the interpolating points in any order; i.e.,

$$P_k(x) = f_k + (x-x_k)f[x_k, x_{k-1}] + (x-x_k)(x-x_{k-1})f[x_k, x_{k-1}, x_{k-2}]$$

$$+ ... + (x-x_k)(x-x_{k-1})...(x-x_1)f[x_k, x_{k-1}, ..., x_1, x_0] .$$

We illustrate with the following example: $f(1) = 1$, $f(2) = 6$, $f(3) = 16$, $f(4) = 35$.

$$P_3(x) = f_0 + (x-x_0)f[x_0, x_1] + (x-x_0)(x-x_1)f[x_0, x_1, x_2]$$

$$+ (x-x_0)(x-x_1)(x-x_2)f[x_0, x_1, x_2, x_3]$$

$$= 1 + x \cdot \frac{5}{2} + x(x-2) \cdot \frac{5}{2} + x(x-2)(x-3) \cdot \frac{1}{2}$$

$$= \frac{1}{2}x^3 + \frac{1}{2}x + 1 .$$

Table 1.1. Divided Difference Table

i	x_i	f_i	$f[x_i, x_{i+1}]$	$f[x_i, x_{i+1}, x_{i+2}]$	$f[x_i, x_{i+1}, x_{i+2}, x_{i+3}]$
0	0	1			
			$\dfrac{5}{2}$		
1	2	6		$\dfrac{5}{2}$	
			10		$\dfrac{1}{2}$
2	3	16		$\dfrac{9}{2}$	
			19		
3	4	35			

Alternatively, we can express

$$P_3(x) = f_3 + (x-x_3)f[x_3, x_2] + (x-x_3)(x-x_2)f[x_3, x_2, x_1]$$

$$+ (x-x_3)(x-x_2)(x-x_1)f[x_3, x_2, x_1, x_0]$$

$$= 35 + (x-4) \cdot 19 + (x-4)(x-3) \cdot \frac{9}{2}$$

$$+ (x-4)(x-3)(x-2) \cdot \frac{1}{2}$$

$$= \frac{1}{2}x^3 + \frac{1}{2}x + 1 .$$

We state the following, useful lemma:

Lemma 1.2.

$$\nabla^r f_k = r!\, h^r f[x_k, x_{k-1}, ..., x_{k-r}] \tag{1.2.18}$$

for any uniformly spaced points $\{x_j\}$ with interval h.

Proof. The proof follows readily by induction, using Relation (1.1.2) and some algebraic manipulations. ∎

The adoption of Lemma 1.2 in the divided difference formula (1.2.12), assuming evenly spaced meshpoints $\{x_k, x_{k-1}, ..., x_1, x_0\}$, yields the following Newton backward interpolation formula:

$$P_k(x) = f_k + (x-x_k)\frac{\nabla f_k}{h} + (x-x_k)(x-x_{k-1})\frac{\nabla^2 f_k}{2!h^2}$$

(1.2.19)

$$+ \dots + (x-x_k)(x-x_{k-1})\dots(x-x_1)\frac{\nabla^k f_k}{k!h^k} .$$

Polynomials $P_k(x)$ of degree k are commonly represented as a linear combination of monomial bases

$$m_i(x) = x^i , \quad i = 0(1)k ,$$

(1.2.20)

which span R^{k+1}; i.e.,

$$P_k(x) = \sum_{i=0}^{k} c_i m_i(x)$$

(1.2.21)

for some real constants c_i, $i = 0(1)k$.

With the $(k+1)$-vectors

$$m(x) = (m_0(x), \dots, m_k(x))^T$$

(1.2.22)

and $c = (c_0, c_1, \dots, c_k)^T$, (1.2.21) can be compactly expressed as

$$P_k(x) = c^T m(x) .$$

(1.2.23)

Jackson and Sedgwick (1977) suggested two other bases in addition to monomial bases (1.2.22):

(i)   Lagrange bases given by

$$l(x) = (l_0(x), l_1(x), \dots, l_k(x))^T ,$$

(1.2.24)

whose components are specified by (1.2.2), and

(ii)   the Newton bases given by

$$\eta(x) = (\eta_0(x), \eta_1(x), \dots, \eta_k(x))^T ,$$

(1.2.25)

where $\eta_i(x)$ is a polynomial of degree i and specified as

$$\eta_0(x) = 1 ,$$

$$\eta_i(x) = \prod_{j=0}^{i-1} (x-x_j) , \; i \geq 1 .$$

(1.2.26)

The Lagrange interpolation (1.2.1) is now reformulated as

$$P_k(x) = f^T l(x) ,$$

(1.2.27)

while the Newton interpolating polynomial (1.2.12) is expressed as

$$P_k(x) = f^T R \eta(x) , \qquad (1.2.28)$$

where R is the coordinate transformation between the Newton and Lagrange bases, thus combining what was previously two separate formulas into one.

R^T acts as the divided difference operator if (1.2.28) is computed as $(f^T R) \eta(x)$, while R acts as a coordinate transformation when (1.2.28) is computed as $f^T (R\eta(x))$. The latter will be appropriate if several sets of data f are to be interpolated at the same set of points, i.e., with the same Lagrange bases.

Jackson and Sedgwick (1977) established the following coordinate transformations from the three bases $\eta(x)$, $l(x)$, and $m(x)$:

$$l(x) = R\eta(x) ,$$

$$\eta(x) = Lm(x) , \qquad (1.2.29)$$

$$m(x) = Vl(x) ,$$

for some upper triangular matrix R, lower triangular matrix L, and the Vandermonde matrix V

$$V = (V_{ij}) = (x_j^i) , \quad 0 \le i,j \le k .$$

It was established that the matrices representing the coordinate transformation between monomial, Lagrange, and Newton bases are the triangular factors of a Vandermonde matrix.

1.3. Finite Difference Equations

A k-th order difference equation can be written as

$$\rho(E)y_n = \phi_n , \qquad (1.3.1)$$

with y_r, $r = 0(1)k-1$ specified. E is the shift operator defined by (1.1.1), and $\rho(E)$ is a polynomial of degree k given as

$$\rho(E) = \sum_{j=0}^{k} \alpha_j E^j , \qquad (1.3.2)$$

where α_j, $j = 0(1)k$ are real constants with $\alpha_0 \ne 0$, $\alpha_k \ne 0$. The roots of ρ are denoted by $r_1, \dots, r_k$.

In the analysis of the numerical method for the initial value problem (IVP), we often encounter the case $\phi_n = \phi$, a constant, and, hence, a particular solution $y_{n,p}$ to (1.3.1) is

$$y_{n,p} = \frac{\phi}{\rho(1)} , \quad \rho(1) \neq 0 . \qquad (1.3.3)$$

The homogeneous part of (1.3.1) is

$$\rho(E)y_n = 0 ,$$

which can be factorized as

$$\prod_{j=1}^{k} (E - r_j)y_n = 0 . \qquad (1.3.4)$$

Equation (1.3.4) suggests k first order difference equations

$$(E - r_j)y_n = 0 \text{ for } j = 1(1)k .$$

This implies

$$y_{n+1} = r_j y_n ,$$

which is solved to give

$$y_{n,j} = r_j^n y_0 , \quad j = 1(1)k \qquad (1.3.5)$$

for some constant y_0.

If the factors of (1.3.4) are distinct, then the k solutions given by (1.3.5) are independent and the solution to the homogeneous Equation (1.3.4) is

$$y_{n,\mathrm{H}} = \sum_{j=1}^{k} \beta_j r_j^n . \qquad (1.3.6)$$

The general solution to (1.3.1), then, is the sum of the solution to the homogeneous equation (1.3.4) and the particular solution (1.3.3); i.e.,

$$y_n = \sum_{j=1}^{k} \beta_j r_j^n + \frac{\phi}{\rho(1)} . \qquad (1.3.7)$$

The constants β_j, $j = 1(1)k$ are normally obtained from the k initial conditions, which result in a set of k linear equations in the unknown vector β:

$$X\beta = y , \qquad (1.3.8)$$

where X is a $k \times k$ matrix whose entries are

$$X_{ij} = r_i^j , \quad i = 1(1)k , \quad j = 0(1)k - 1 , \qquad (1.3.9)$$

and β, y are k-tuples

$$\beta = (\beta_1, \beta_2, ..., \beta_k)^T , \quad y = (y_0, y_1, ..., y_{k-1})^T . \qquad (1.3.10)$$

Clearly, X is a Vandermonde matrix, and so the equations are uniquely solvable if ρ has k different roots.

It may happen that (1.3.4) has a multiple root. Suppose $r_1 = r_2 = r_3 = ... = r_s$, $s \le k$ in (1.3.4). The Solution (1.3.6) to the homogeneous equation is replaced by

$$y_{n,H} = (\beta_1 + n\beta_2 + n(n-1)\beta_3 + ... + n(n-1)...(n-s+2)\beta_s)r_1^n$$

(1.3.11)

$$+ \sum_{j=}^{k} \beta_j r_j^n .$$

Example. Solve the fourth order difference equation

$$y_{n+4} + 2y_{n+3} + 2y_{n+2} + 2y_{n+1} + y_n = 8 ,$$

(1.3.12)

whose initial conditions are

$$y_0 = 4, \ y_1 = -4, \ y_2 = 4, \ y_3 = -4 .$$

(1.3.13)

The characteristic polynomial of the homogeneous part of (1.3.12) is

$$\rho(E) = E^4 + 2E^3 + 2E^2 + 2E + 1$$

$$= (E+1)^2(E-i)(E+i) ,$$

with the solution $y_{n,H}$ of the homogeneous equation obtained from (1.3.11) as

$$y_{n,H} = (\beta_1 + n\beta_2)(-1)^n + \beta_3 i^n + \beta_4(-i)^n .$$

A particular solution to (1.3.12) is

$$y_{n,p} = \frac{8}{\rho(1)} = 1 ,$$

leading to a general solution

$$y_n = (\beta_1 + n\beta_2)(-1)^n + \beta_3 i^n + \beta_4(-i)^n + 1 .$$

(1.3.14)

Adopting the initial conditions (1.3.13) in (1.3.14) gives the linear system (1.3.8) with

$$X = \begin{pmatrix} 1 & 0 & 1 & 1 \\ -1 & -1 & i & -i \\ 1 & 2 & -1 & -1 \\ -1 & -3 & -i & i \end{pmatrix} ,$$

$$y = (4, -4, 4, -4)^T ,$$

whose solution β is

$$\beta = (1, 2, 1+i, 1-i)^T .$$

This, in (1.3.14), gives the solution to (1.3.12) as

$$y_n = (1+2n)(-1)^n + 2(-1)^{[(n+1)/2]} + 1 . \tag{1.3.15}$$

1.4. Linear Systems with Constant Coefficients

The general nonlinear initial value problem

$$y\prime = f(x, y), \ y(a) = \eta , \tag{1.4.1}$$

$x \, \varepsilon \, R$, y, f, $\eta \, \varepsilon \, R^m$ may be satisfactorily represented locally by the linear system

$$y\prime = A(x)y + d(x) , \ y(a) = \eta , \tag{1.4.2}$$

where $A = (\partial f / \partial y)$ is the Jacobian matrix. In control theory, the stability of (1.4.1) in the Lyapunov sense is often predicted by the stability properties of (1.4.2) (cf. Barnett 1975).

The homogeneous part of (1.4.2) is

$$y\prime = A(x)y . \tag{1.4.3}$$

If λ_i, $i = 1(1)m$ are the eigenvalues of A (assumed to be distinct) and c_i, $i = 1(1)m$ the corresponding eigenvectors, we have that

$$Ac_i = \lambda_i c_i , \quad i = 1(1)m . \tag{1.4.4}$$

Since the λ_i are distinct, then the m eigenvectors span C^m, and the elements of the fundamental system

$$y_{i,H} = e^{\lambda_i x} c_i , \quad i = 1(1)m \tag{1.4.5}$$

are linearly independent. Hence, a general solution of (1.4.3) is given by

$$y_H = \sum_{i=1}^{m} \alpha_i e^{\lambda_i x} c_i , \tag{1.4.6}$$

where α_i, $i = 1(1)m$ are arbitrary constants whose numerical values can be obtained from the initial conditions $y(a) = \eta$.

With y_p as a particular solution to (1.4.2), a general solution to (1.4.2) is

$$y(x) = \sum_{i=1}^{m} \alpha_i e^{\lambda_i x} c_i + y_p(x) . \tag{1.4.7}$$

If the solution fulfills the initial conditions, the coefficients α_i, $i = 1(1)m$ are determined from the following system of linear equations:

$$\sum_{i=1}^{m} \alpha_i e^{\lambda_i a} c_i = \eta - y_p(a) . \qquad (1.4.8)$$

For a general discussion on this topic, interested readers should refer to Sanchez (1968), Barnett (1975), and Hirsch and Smale (1974).

Example. The initial value problem $y'' + y = x + e^{-2x}$, $y(1) = 1$, $y'(1) = 0$ has a particular solution $y_p(x) = x + 1/5e^{-2x}$. Construct the equivalent first order system and find its general solution.

Substitute

$$y_1 = y , y_1' = y_2$$

in the second order IVP to get

$$y_1' = y_2 , y_1(1) = 1 ,$$

$$y_2' = -y_1 + x + e^{-2x} , y_2(1) = 0 .$$

This is equivalent to the matrix form (1.4.2) with

$$A = \begin{bmatrix} 0 & 1 \\ -1 & 0 \end{bmatrix} , d(x) = (1, x + e^{-2x})^T . \qquad (1.4.9)$$

The eigenvalues of A are $\lambda_1 = +i$, $\lambda_2 = -i$, whose eigenvectors are scalar multiples of the vectors

$$c_1 = (1, i)^T , c_2 = (1, -i)^T .$$

With $y_p(x) = x + 1/5e^{-2x}$ as a particular solution of the second order difference equation, a particular solution of the corresponding first order system (1.4.9) is

$$y_p(x) = (x + \frac{1}{5}e^{-2x}, 1 - \frac{2}{5}e^{-2x}) .$$

Hence, from (1.4.7), the general solution of the given problem is

$$y(x) = \alpha_1 \begin{bmatrix} 1 \\ i \end{bmatrix} e^{ix} + \alpha_2 \begin{bmatrix} 1 \\ -i \end{bmatrix} e^{-ix} + \begin{bmatrix} x + \frac{1}{5}e^{-2x} \\ 1 - \frac{2}{5}e^{-2x} \end{bmatrix} . \qquad (1.4.10)$$

Corresponding to (1.4.8) is the linear system

$$\alpha_1 + \alpha_2 = \frac{4}{5} ,$$

$$i\alpha_1 - i\alpha_2 = -\frac{3}{5} . \tag{1.4.11}$$

$\alpha_1 = 4+3i/10$, $\alpha_2 = 4-3i/10$ satisfy the pair of Equations (1.4.11).

Hence, the solution of the IVP is

$$y_1(x) = \frac{4}{5}\cos x - \frac{3}{5}\sin x + x + \frac{1}{5}e^{-2x} \tag{1.4.12}$$

and

$$y_2(x) = -\frac{4}{5}\sin x - \frac{3}{5}\cos x + 1 - \frac{2}{5}e^{-2x} .$$

1.5. Distribution of Roots of Polynomials

In the stability analysis of discretization methods for IVPs in ODEs, one is often faced with the problem of determining the distribution of the roots of a polynomial of the form

$$P_k(r, z) = \sum_{j=0}^{k} a_j(z)r^j , \tag{1.5.1}$$

$z \, \varepsilon \, C$ and $a_k(z) \neq 0$, $a_0(z) \neq 0$.

Since it is not the exact values of the roots that are required, but the location of the roots relative to the unit disc, the Schur criterion is a very appropriate tool (cf. Miller 1971). Setting the righthand side of (1.5.1) to zero leads to an algebraic equation with solutions that are global analytic vectors of functions that "live" on a Riemann surface. This approach makes stability analysis more understandable and intuitive; besides, the second Dahlquist barrier theorem is not, surprisingly, any longer.

Definition 1.1. The polynomial $P_k(r, z)$ is said to be of the type (q_1, q_2, q_3) provided it has (counting multiplicities) q_1 zeros inside the unit disc D, q_2 zeros on the boundary S of D, and q_3 zeros outside S with $k = q_1 + q_2 + q_3$.

From (1.5.1), we derive the polynomial $P_k^*(r, z)$ as

$$P_k^*(r, z) = \sum_{j=0}^{k} a_j^*(z)r^{k-j} , \tag{1.5.2}$$

where $a_j^*(z)$ denotes the complex conjugate of $a_j(z)$. P_k being of the type (q_1, q_2, q_3) implies P_k^* is of the type (q_3, q_2, q_1).

Equations (1.5.1) and (1.5.2) can be combined to obtain the following polynomial of degree $k-1$:

$$P_{k-1}(r, z) = [P_k^*(1, z)P_k(r, z) - P_k^*(r, z)P_k(1, z)]/r . \tag{1.5.3}$$

Definition 1.2. P_k is said to be self-inversive if it has the same zeros as P_k^*.

P_k must be of the type $(q, k-2q, q)$ and $P_{k-1} \equiv 0$ since the coefficients are necessarily symmetric; i.e. $a_j(z) = a_{k-j}(z), j = 0(1)\bar{k}, \bar{k} = [k/2]$.

Theorem 1.4 (Miller 1971). *Suppose* $P_k(r, z)$ *is a polynomial of degree* k *such that* $|P_k(1, z)| \neq |P_k^*(1, z)|$. *Then* P_k *is of the type* (q_1, q_2, q_3) *iff* P_{k-1} *is of the type* (q_1-1, q_2, q_3) *when* $|P_k^*(1, z)| > |P_k(1, z)|$, *and of the type* (q_3-1, q_2, q_1) *if* $|P_k^*(1, z)| < |P_k(1, z)|$.

The reduction process (1.5.3) can be generalized with the following set of polynomials:

$$P_{j-1}(r, z) = [P_j^*(1, z)P_j(r, z) - P_j^*(r, z)P_j(1, z)]/r \qquad (1.5.4)$$

for $j = k(-1)1$.

Definition 1.3. $P_k(r, z)$ is a Schur polynomial if it is of the type $(k, 0, 0)$.

Theorem 1.5. P_k *is a Schur polynomial iff*

$$|P_k^*(1, z)| > |P_k(1, z)| , \qquad (1.5.5)$$

and P_{k-1} *is a Schur polynomial.*

Example. Locate the roots of the polynomial

$$P_3(r, z) = r^3 - r^2 - \frac{z}{24}(9r^3+19r^2-5r+1) , z = -1 \qquad (1.5.6)$$

with respect to the unit circle.

$$P_3(r, -1) = \frac{33}{24}r^3 - \frac{5}{24}r^2 - \frac{5}{24}r + \frac{1}{24} ,$$

$$P_3(1, -1) = \frac{1}{24} ,$$

$$P_3^*(r, -1) = \frac{1}{24}r^3 - \frac{5}{24}r^2 - \frac{5}{24}r + \frac{33}{24} ,$$

$$P_3^*(1, -1) = \frac{33}{24} ,$$

$$P_2(r, -1) = \frac{1}{18}(34r^2-5r-5) ,$$

$$P_2(1, -1) = -\frac{5}{18} ,$$

$$P_2^*(r, -1) = \frac{1}{18}(-5r^2 - 5r + 34),$$

$$P_2^*(1, -1) = \frac{34}{18},$$

$$P_1(r, -1) = \frac{1}{324}(1181r - 145),$$

$$P_1(1, -1) = -\frac{145}{324},$$

$$P_1^*(1, -1) = \frac{1181}{324}.$$

From Theorems 1.4 and 1.5, we observe that $P_3(r, -1) \equiv (3, 0, 0)$.

Exercise. Locate the roots of (1.5.6) with respect to the unit circle for the following cases:

(i) $z = -5$,

(ii) $z = -7$,

(iii) $z = +1$,

(iv) $z = 0$,

(v) $z = -6$.

Miller (1971) extended the Schur Theorems 1.4 and 1.5 to locate the roots of (1.5.1) relative to any directed line segment L in the complex plane. $P_k(r, z) \equiv (q_1, q_2, q_3)$ relative to L if it has q_1 zeros to the left of L, q_2 on L, and q_3 to the right of L.

Lambert (1973a), Sand (1979), and Sand and Østerby (1979) discussed alternative approaches based on the Routh-Hurwitz criterion and Root Locus method. Wanner et al. (1978) adopted a natural and geometric approach of "order stars" to establish some important stability theorems.

1.6. First Integral Mean Value Theorem

Theorem 1.6. *Let $\phi(x)$ be continuous in $[a, b]$ and $f(x)$ integrable and of constant sign in $[a, b]$. Then*

$$\int_a^b \phi(x) f(x)dx = \phi(\xi) \int_a^b f(x)dx \tag{1.6.1}$$

for some $\xi \in (a, b)$.

The proof of this theorem can be found in Apostle (1957) and Bartle (1975).

1.7. Common Norms in ODEs

(i)   *The Maximum Norm* is defined as

$$v_\infty = \max_i |v_i| \, , \tag{1.7.1}$$

where v is an n-vector.

(ii) *The Euclidean Norm* is defined as

$$v_2 = (\sum_{i=1}^{n} v_i^2)^{1/2} \, , \tag{1.7.2}$$

and

(iii) the *Root Mean Square Norm (RMS)* is defined as

$$v_{\text{RMS}} = (\frac{1}{n} \sum_{i=1}^{n} v_i^2)^{1/2} \, . \tag{1.7.3}$$

Both the Euclidean norm and the Root Mean Square norm yield smoother behavior than the maximum norm but are susceptible to underflow, particularly for stiff initial value problems.

In the case of the Maximum norm, it is natural to adopt the subordinate matrix norm, but this is impractical with the Euclidean norm. A compatible Frobenius matrix norm could be used with the risk that this can be much larger than M_2 (where M denotes a matrix of order n).

A possibility is the adoption of the *One-Norm*

$$v_1 = \sum_{i=1}^{n} |v_i| \, , \tag{1.7.4}$$

or the *Mean Magnitude Norm (MM)*

$$v_{\text{MM}} = \frac{1}{n} \sum_{i=1}^{n} |v_i| \, , \tag{1.7.5}$$

which are both insulated from the scaling problems associated with the Euclidean norm and the lack of smoothness associated with the Maximum norm.

We shall, in the course of this monograph, discuss the numerical treatment of some difficult IVPs that normally arise from the magnitude of the Jacobian matrix. Hence, we shall briefly discuss some matrix norms. Golub and Sloan (1983) will be a good reference.

The subordinate matrix norm for the one-norm is

$$M_1 = \max_j \sum_{i=1}^{n} |M_{ij}| \, , \tag{1.7.6}$$

which is easily computable and is well-suited to all classes of initial value problems.

Both the Root Mean Square norm and the Mean Magnitude norm are liberal in error control for stiff problems, while the Maximum norms are more suitable for nonstiff problems because of their uniformity. The One-Norm and the Euclidean Norm still remain favorite in ODE codes.

Other topics of interest that may enable readers to grasp easily the contents of this book include solution of linear equations (cf. Atkinson 1978) and the solution of nonlinear equations (cf. Ortega and Rheinboldt 1970).

The three devices generate series of solutions whose convergence cannot be guaranteed, and even when the series do converge, the rate of convergence may be rather low.

Corliss and Chang (1982) demonstrated the effectiveness of the Taylor series on nonstiff test problems, i.e., problems with small Lipschitz constants (cf. Hull et al. 1972). They asserted that the Taylor series method is most attractive for small systems and for stringent accuracy tolerance.

Miller (1946) was the first to use recurrence schemes in computing the Airy Integral for the British Association Mathematical Table. Moore (1966, 1975), Barton (1971a,b), Gear (1971c), Chang (1974), Chang and Corliss (1980), Chang et al. (1979), Corliss and Lowery (1977), Norman (1973, 1976), Rall (1981), and Halin (1976, 1983) proposed algorithms or translator programs to accept differential equations as input and to produce object codes for solving a given differential system. Barton (1980) and Chang (1986) examined the possibility of using modified Taylor series methods to cope with stiff IVPs (i.e., inherently stable differential systems with large Lipschitz constants), highlighting the fact that only infinite Taylor series methods can attain A-stability.

(b) The second class of numerical methods for (2.1.4) is the expansion methods whereby the theoretical solution $y(x)$ is approximated by the first few terms of an expansion in orthogonal functions:

$$y_N(x) = \sum_{i=1}^{N} \alpha_i H_i(x) . \tag{2.1.8}$$

The expansion coefficients α_i can depend on N and possibly on $H_i(x)$. The functions $H_i(x)$ may be defined over the entire integration interval (a, b), as in the case of the Chebyshev polynomial (Fox and Parker 1968), or can be defined over short intervals, as with splines. There is a close relationship between the two methods.

(c) Finally, and most important to this text, are the discrete variable methods whereby numerical approximations are obtained at some specified points in the integration interval. Some of these methods can be described as a more generally applicable version of Taylor series. This approach is more universally applicable, since, in general, it is more amenable to computer implementation. It is not surprising, then, that the advent of the computer era within the last five decades has greatly accelerated the pace of development of algorithms based on this approach.

This text will be essentially dedicated to the theory, development, and implementation of both currently existing and new algorithms based on the discrete variable approach.

Other approaches like the collocation methods do possess their own merits.

Let ε denote a specified allowable error tolerance.

A discrete point set

$$\{x_n \,|\, x_0 = a, \; x_{n+1} = x_n + h_n, \quad n = 0(1)N - 1, \; x_N = b\}$$

is defined on a closed and finite interval [a, b] of the real line, where the discretization parameter ε dictates the choice of meshsize h_n.

Definition 2.1. An integration algorithm is a complete and unambiguous logical sequence of operations which transforms an allowable input data vector $(a, b, y_0, f, hprin, \varepsilon)$ into allowable output data vectors {a set of vectors y's at points $(x_0 + hprin, \; x_0 + 2\,hprin, etc.)$}, where $hprin$ is a specified printing interval, $y(x, h)$ is the numerical estimate to the theoretical solution $y(x)$, and le_n is the estimated local or global error (usually the local truncation error) at an output point x.

The acceptability of an integration algorithm depends on the following factors:

(a) The degree of compatibility of the numerical solution at the specified output points. Is $y(b, h)$ sufficiently close to $y(b)$? Or is $|y(x, h) - y(x)| < \varepsilon$ at a printing point?

(b) The amount of computer time required or the volume of computation required to integrate from $x = a$ to $x = b$. This, to a large extent, will depend on the number of evaluations of the derivative $f(x, y)$, which, in turn, depends on the acceptable meshsizes.

(c) Reasonable degree of reliability (i.e., ability to achieve desired accuracy (Sedgwick 1973)).

In the last two decades, significant improvements in efficiency, convenience, and reliability have been achieved by numerical integration codes for the initial value problems (2.1.4).

2.2. Existence of Solution, Numerical Approach

It may happen that the IVP (2.1.4) has no solution, or a solution may exist without it being unique unless some more analytic properties are incorporated into the integration algorithm.

Figure 2.1

It is then worthwhile to first examine if the IVP possesses certain important properties.

Definition 2.2. The IVP (2.1.4) is said to satisfy a Lipschitz condition of order one w.r.t. y if there exists a finite constant L such that whenever two points (x, y) and (x, z) are in the solution space

$$R^{m+1} = [a, b] \times \{y \mid y_\infty \leq \tau < \infty\} , \tag{2.2.1}$$

the following relationship holds:

$$f(x, y) - f(x, z)_\infty \leq L y - z_\infty , \tag{2.2.2a}$$

where $\cdot_\infty$ is the maximum norm defined as

$$y_\infty = \max_{1 \leq r \leq m} |y_r| . \tag{2.2.3}$$

Henceforth, the occurrence of $\cdot$ is presumed to imply (2.2.3), which we adopt simply because of its ease of computation.

The classical Lipschitz condition (2.2.2a) is often inadequate in an attempt to control the qualitative behavior of the exact solution to (2.1.4), as it is incapable of differentiating between the test problems $y\prime = \lambda y$ or $y\prime = -\lambda y$. A possible remedy is to adopt the one-sided Lipschitz condition:

$$<f(x, y) - f(x, z), y - z> \leq V(x)y - z^2 , \tag{2.2.2b}$$

where $V(x)$ (which can be negative) is the one-sided Lipschitz "constant" of f with respect to y. Equation (2.2.2b) implies (2.2.2a), but the converse is not true in general.

We can now state the existence and uniqueness theorem, whose proof can be found in many textbooks on differential equations, e.g., Ince (1956), Henrici (1962), Pontryagin (1962), and Coddington and Levinson (1955).

Theorem 2.1. *Let $f(x, y)$ be defined and continuous on R given by (2.2.1) and, in addition, satisfy Inequality (2.2.2). Then the initial value problem (2.1.4) has a unique solution in R.*

Since the Lipschitz constant L is not readily available, it suffices to check if f_y (the partial derivative of $f(x, y)$ with respect to y) is defined and continuous in R. In other words, it is expected that all the first partial derivatives of f are bounded by a constant K; i.e., $|\partial f_i/\partial y_j| \leq K$ for all $(x, y) \, \varepsilon \, R$. For instance, the IVP

$$y\prime = \sqrt{|y|} , \quad y(0) = -1 , \tag{2.2.4}$$

whose integral curves are displayed in Figure 2.2, has no unique solution, as f_y is undefined at $x = 2$.

The conditions of the existence theorem can be weakened by demanding that they hold piecewise in the interval $a \leq x \leq b$ of interest.

The conventional discrete variable methods fall into two distinct categories:

(a) The one-step or Runge-Kutta methods (R-K), which are essentially substitution methods (Chapters 3, 4, 8 and 9).

(b) The Linear Multistep methods (LMM), which are basically polynomial interpolation schemes (Chapters 5, 8 and 9), and One-leg Multistep methods (Chapters 8 and 10).

Methods which may fall into either of the two distinct groups include:

(i) The Hybrid methods, which combine the characteristics of both (a) and (b) (Chapter 10).

(ii) The Extrapolation methods are discretization methods whereby successive terms in the asymptotic error expansion in h or h^2 are annihilated (Chapters 7 and 9). Shampine (1984) showed that the polynomial extrapolation applied to the modified midpoint rule is equivalent to a variable order Runge-Kutta methods.

(iii) Nonlinear Multistep methods, which are designed for "unconventional" problems (Chapter 6). Apparently, such schemes subject themselves to a limited degree of analysis. Some of these schemes have proved to be quite effective for dealing with the special initial value problems discussed in the next section.

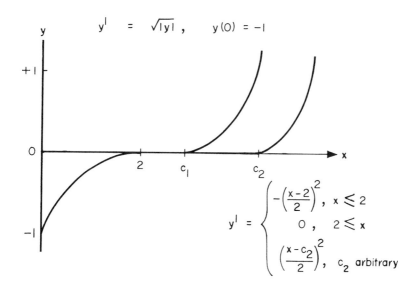

Figure 2.2: Consequence of Violation of Theorem 2.1. No Unique Solution

2.3. Special IVPs

The conventional numerical integrators are, in general, formulated on the basis of polynomial interpolation, with the tacit assumption that the IVP (2.1.4) satisfies the hypothesis of Theorem 2.1. Consequently, such algorithms perform poorly when they are applied to IVPs that violate the hypothesis of Theorem 2.1. Examples of such problems shall be discussed next.

First, consider the harmless looking IVP

$$y' = 1 + y^2 , \quad y(0) = 1 \tag{2.3.1a}$$

in the interval $0 \leq x \leq \pi/4$. The theoretical solution to (2.3.1a) is $y(x) = \tan(x+\pi/4)$ which is unbounded at $x = \pi/4$. The singularity in Problem (2.3.1a) can only be bypassed by analytic continuation in the complex plane.

The conventional numerical integrators are thus inefficient, as they merely track the solution which "explodes" in the neighborhood of this singularity, as the IVP (2.3.1a) violates the hypothesis of Theorem 2.1; i.e., its Lipschitz constant L or f_y is unbounded near $x = \pi/4$. The problem is compounded by the fact that there may be no clue as to the location of these singularities, particularly if $f(x, y)$ is nonlinear.

Another class of problems that are difficult to solve by conventional numerical integrators is of the form:

$$\begin{aligned} y' &= 0, \quad x < 0 \\ &= x^6, \quad x \geq 0 \end{aligned}, \quad y(-1) = 0 , \tag{2.3.1b}$$

whose theoretical solution is

$$\begin{aligned} y(x) &= 0 , \quad x < 0 , \\ &= \frac{x^7}{7} , \quad x \geq 0 . \end{aligned}$$

Shampine created Problem (2.3.1b) to expose a defect in stepsize selection scheme of DIFSUB, and it has been used subsequently to reveal what is happening on other codes.

Such problems can be readily handled with the one-step scheme by ensuring that the point of discontinuity is a meshpoint. The subroutine that evaluates the derivatives contains a switch that alerts the integrator when a discontinuity is to be overstepped. The three essential problems posed by (2.3.1b) are

(a)   detecting a discontinuity,

(b)   locating the point of discontinuity, and

(c)   restarting the integration beyond the point of discontinuity.

Carver (1977) in the code FORSIM provided options to locate discontinuity; Gear and Østerby (1981) investigated the same problem for nonstiff problems, while Hindmarsh (1981) incorporated a root finder in the code LSODAR. Gear (1980d) proposed a suitable Runge-Kutta step to generate enough information for a four-step multistep scheme to continue accurately beyond discontinuity. Halin (1983) used an automatic Taylor series method to achieve the same objective. Other authors who advocated the use of the switching function approach to cope with discontinuity include O'Regan (1970), Hay et al. (1974), Evans and Fatunla (1975), Carver (1977, 1978), Halin (1976, 1983), and Gear and Østerby (1984). The Shampine and Gordon (1975) Adams code ODE/STEP, INTRP has a number of provisions specifically for this difficulty, which require no actions by the user such as a rootfinder. Enright et al. (1986) used interpolation in R-K formulas to develop efficient and reliable automatic technique for discontinuous IVPs.

Another important problem area includes IVPs in the form

$$y\prime = \lambda(y - E(x)) + E\prime(x), \ y(0) = 1.0 \ ,$$

$$E(x) = A - (A + x)e^{-x} \ , \qquad\qquad (2.3.2)$$

$$\lambda = -2000, \ A = 10, \quad 0 \le x \le 100 \ .$$

The theoretical solution to Problem (2.3.2) is given by

$$y(x) = E(x) + 10e^{-2000x}.$$

$E(x)$ is slowly varying in the interval of integration, while the second component decays rapidly in the "transient phase." The IVPs with such properties are said to be stiff, and the application of the conventional discretization algorithms to such problems will require the adoption of an intolerably small meshsize in the entire integration interval. This will result in exorbitant computing time, and, besides, there is a tendency for accumulation of roundoff errors in nonstiff components.

Robust algorithms like Gear's DIFSUB (1971a), Hindmarsh's LSODE (1980), DEPAC (Shampine and Watts, 1979), Enright's Second Derivative Formulas (1974a), and Skeel and Kong's Blended-DIFSUB (1977) have considerably eased the situation. Other efficient codes include METAN1 (Bader and Deuflhard 1983), which is based on polynomial extrapolation applied to the semi-implicit midpoint rule; STRIDE (Burrage, Butcher, and Chipman 1980), based on the singly implicit Runge-Kutta process originally developed by Burrage (1978a); STINT, based on variable step, variable order code, which uses cyclic composite multistep methods (Tendler et al. 1978a,b; Iserles 1984; Albrecht 1978, 1985, 1986, 1987; Tischer and Sacks-Davis 1983; Tischer and Gupta 1983, 1985); the code SIMPLE based on SDIRK (Norsett 1986); and MEBDF, based on Modified Extended Backward Differentiation Formula (Cash 1983c).

The third problem area includes problems like

$$\begin{bmatrix} y_1(x) \\ y_2(x) \end{bmatrix}' = \begin{bmatrix} -.00005 & 100 \\ -100 & -.00005 \end{bmatrix} \begin{bmatrix} y_1 \\ y_2 \end{bmatrix} , \tag{2.3.3}$$

$$y(0) = (0, 1)^T , \quad 0 \le x \le 10\pi ,$$

whose theoretical solution is obtained as

$$\begin{bmatrix} y_1(x) \\ y_2(x) \end{bmatrix} = e^{-0.00005x} \begin{bmatrix} \sin 100x \\ \cos 100x \end{bmatrix} . \tag{2.3.4}$$

The transitory phase for this problem is the entire interval of integration $0 \le x \le 10\pi$, with $50/\pi$ complete oscillations per unit interval.

As in the two previous problems discussed, the conventional integration algorithms cannot adequately cope with Problem (2.3.3), which is considered to be highly oscillatory.

Gear (1980b) discussed automatic procedures for the detection and treatment of stiff and/or oscillatory differential equations.

Efforts will be made in the course of this text to focus attention to these special problem areas, which have attracted a lot of research attention within the last decade.

2.4. Error Propagation, Stability, and Convergence of Discretization Methods

The integral curve (local solution) $y(x; x_n, y_n)$ to the IVP (2.1.4) is the solution of

$$u' = f(x, u), u(x_n) = y_n , \tag{2.4.1}$$

where y_n is an approximation to the exact solution $y(x_n)$.

The local error t_{n+1} is

$$t_{n+1} = y(x_{n+1}; x_n, y_n) - y_{n+1} . \tag{2.4.2}$$

Few codes control the local error committed at every integration step by demanding that

$$t_{n+1} \le h_n \tau_n , \tag{2.4.3}$$

where $h_n = x_{n+1} - x_n$ is the current stepsize and τ_n is the allowable error tolerance, which may depend on the independent variable x_n. Most practical codes, however, replace h_n on the righthand side of (2.4.3) by unity, thus adopting the error per step criterion.

A user is actually interested in the true or global error e_{n+1} specified by

$$e_{n+1} = y(x_{n+1}) - y_{n+1} , \tag{2.4.4}$$

and it is well known that the variational equation

$$e\prime(x) = J(x, \, y\,(x))e(x), \, e\,(a) = \delta \qquad (2.4.5)$$

(where J is the Jacobian matrix associated with the IVP (2.1.4)) says how an error of δ at $x = a$ propagates. The approximate equation (2.4.5) is satisfied by the error-neglecting second order terms (cf. Shampine 1984a, 1985b).

The propagation of errors depends on two factors:

(i)   the local error and

(ii) the nature/stability of the problem as described by (2.4.5).

For instance, if the IVP (2.1.4) is inherently stable (i.e., all the eigenvalues of J have negative real parts), then the local errors may damp out with increasing x; otherwise, the errors will be magnified with increasing x (cf. Lambert 1980).

Bulirsch and Stoer (1966) constructed asymptotic upper and lower bounds on the global errors emanating from extrapolation methods to IVPs. Shampine (1985d) generalized this idea for any one-step method endowed with an asymptotically correct local error estimator.

There exist fundamental obstacles in the direct control of the global error, but in recent years, appreciable progress has been attained in efficient and reliable

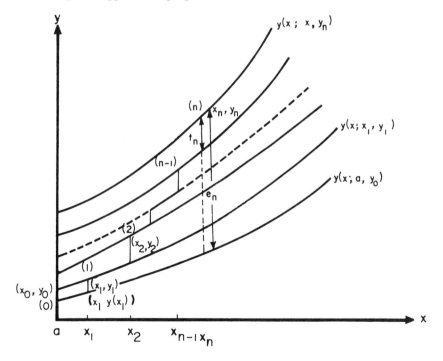

Figure 2.3: Geometrical Representation of Local and Global Error

estimation of the global error (cf. Gear 1974; Stetter 1978, 1979a,b,c, 1981, 1982; Robinson and Prothero 1977, 1980; Shampine and Watts 1976a, 1977; Shniad 1980; Shampine 1983a; Skeel 1982, 1985, 1986a,b; Dormand et al. 1984;, Shampine and Baca 1985; and Hairer and Lubich 1984). Quite often, the calculation is repeated with a different tolerance, and the level of accuracy is inferred from the two solutions. Shampine (1980) warned that this could be misleading. Gladwell has a code D02BDF based on Merson method in the NAG Library that produces global error estimates and performs stiffness checks. The global error estimate is calculated using the two-and-one technique (cf. Shampine and Wattts 1976a).

The local discretization errors and the roundoff errors constitute a sequence of perturbations that shift computed solutions to neighboring integral curves, as illustrated in Figure 2.3.

The stability of a discretization method for (2.1.4) demands that, provided the starting global error $e_0 \neq 0$, the ultimate global error e_n should be bounded; i.e., a finite constant K exists such that

$$e_n < K e_0 . \tag{2.4.6}$$

Similarly, for a discretization method for (2.1.4) to be convergent, it is necessary that the numerical approximation y_n be constrained to be as close to $y(x_n, a, y_0)$ as may be desired with a choice of the appropriate meshsizes h_n; i.e.,

$$\lim_{h_n \to 0} y_n = y(x_n; a, y_0) . \tag{2.4.7}$$

3. THEORY OF ONE-STEP METHODS

3.1. General Theory of One-Step Methods

The conventional one-step numerical integrator for the IVP (2.1.4) can be described as follows:

$$y_{n+1} = y_n + h_n \phi(x_n, y_n; h_n) , \tag{3.1.1}$$

where $\phi(x, y; h)$ is the increment function and h_n is the meshsize adopted in the subinterval $[x_n, x_{n+1}]$. For the sake of convenience and easy analysis, h_n shall be considered fixed (i.e., $h_n = h$, $n \geq 0$). The implementation of a typical one-step scheme is illustrated in Figure 3.1.

Definition 3.1. The integration formula (3.1.1) is said to be consistent with the IVP (2.1.4) provided the increment function $\phi(x, y; h)$ satisfies the following relationship:

$$\phi(x, y; 0) = f(x, y) . \tag{3.1.2}$$

The consistency of a one-step numerical integrator ensures that the scheme is at least of order one. Basically, consistency of a formula ensures that the method approximates the ODE in some sense. The essence of consistency will become manifest when the notion of order is defined later in Definition 3.3.

Definition 3.2. The local truncation error (l.t.e.) t_{n+1} of the one-step scheme (3.1.1) is given by

$$t_{n+1} = y(x_{n+1}) - y(x_n) - h\phi(x_n, y(x_n); h) , \tag{3.1.3}$$

where $y(x)$ is the true solution to (2.1.4).

The l.t.e., as defined in (3.1.3), is simply the amount by which the true solution of the IVP (2.1.4) fails to satisfy the first order difference equation

31

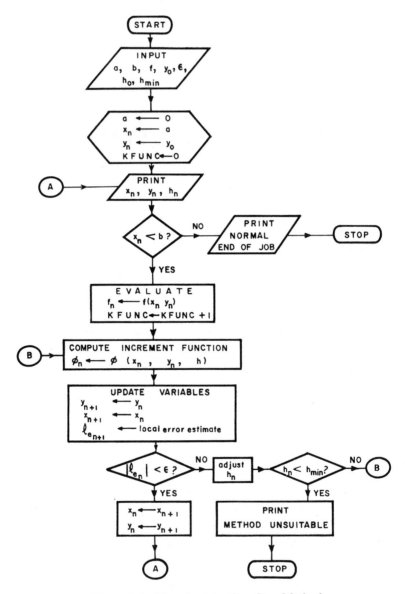

Figure 3.1: Flowchart for One-Step Method

(3.1.1) under the simplifying assumption that the previous solutions are exact (i.e., $y_n = y(x_n)$). The finite difference equation is briefly discussed in Section 1.3.

We assume that the theoretical solution $y(x)$ belongs to the class $C^\infty [a,b]$.

Definition 3.3. The integration formula (3.1.1) is said to be of order p if p is the largest positive integer such that the l.t.e. t_{n+1} satisfies

$$t_{n+1} = O(h^{p+1}),$$ (3.1.4)

where $O(h^{p+1})$ implies the existence of finite constants C and $h_0 > 0$ such that

$$t_{n+1} \le Ch^{p+1}$$

for all $h \le h_0$. Consistency normally implies that the order $p \ge 1$.

Definition 3.4. The global error e_{n+1} of (3.1.1) is the difference between the theoretical solution $y(x_{n+1})$ and the numerical solution y_{n+1}; i.e.,

$$e_{n+1} = y_{n+1} - y(x_{n+1}).$$ (3.1.5)

Definition 3.5. The one-step scheme (3.1.1) is considered convergent if, for an arbitrary initial solution vector y_0 and an arbitrary point $x \, \varepsilon[a, b]$, the global error fulfills the following relationship:

$$\lim_{h \to 0} \max_n e_n \to 0,$$ (3.1.6)

provided x is always a meshpoint.

Definition 3.6. The integration scheme (3.1.1) is stable, if for any initial error e_0, there exist constants K and $h_0 > 0$ such that, when (3.1.1) is applied to the IVP (2.1.4) with meshsize $h \, \varepsilon(0, h_0]$, the ultimate error e_n satisfies the following inequality:

$$e_n \le Ke_0, \quad K < \infty.$$ (3.1.7)

A special stability concept is that of absolute stability, which is normally associated with inherently stable IVPs (those IVPs (2.1.4) whose Jacobians $(\partial f/\partial y)$ have eigenvalues with negative real parts (cf., Sanchez 1968, Barnett 1975).

The absolute stability properties of (3.1.1) can be investigated by applying (2.1.1) to the scalar test problem

$$y' = \lambda y$$ (3.1.8)

(λ is a complex constant with negative real part), which yields the following first order difference equation:

$$y_{n+1} = \mu(z)y_n, \quad z = \lambda h.$$ (3.1.9)

Almost invariably, the stability function $\mu(z)$ is either a polynomial or a rational function in z, with the latter a more desirable property. The parameters in (3.1.1) may be chosen to ensure that $\mu(z)$ is a Padé approximation to e^z. This leads us to the following definition:

Let $P_r(z)$ denote the polynomial of degree r in z specified as

$$P_r(z) = 1 + \frac{r}{2r}\, z + \frac{r(r-1)}{(2r)(2r-1)}\, \frac{z^2}{2!}$$

(3.1.10)

$$+\ldots+ \frac{r(r-1)\ldots 1}{(2r)(2r-1)\ldots(r+1)}\, \frac{z^r}{r!}\, .$$

Then the function

$$R_{r,s}(z) = P_r(z)/P_s(-z)$$

(3.1.11a)

is the (r, s) Padé approximation to e^z of order $r + s$. The Function (3.1.11a) is only valid for $r = s$. A general Padé approximation is

$$R_{r,s}(z) = \sum_{k=0}^{r} \binom{r}{k} \frac{(r+s)!}{(r+s-k)!}\, z^k\, .$$

(3.1.11b)

The theory of Padé approximation is very useful in the development and analysis of numerical methods for stiff initial value problems.

In the A-stability analysis of many classes of one-step methods (e.g., R-K processes, multiderivative methods, Rosenbrock methods, or collocation type schemes) for stiff initial value problems, there is always the desire that the stability function $\mu(z)$ be a rational approximation to the exponential function (3.1.11), which is bounded by one in the entire left half plane $\mathrm{Re}(z) \leq 0$.

Definition 3.7. The one-step scheme (3.1.1) is said to be absolutely stable at a point z in the complex plane provided the stability polynomial/function $\mu(z)$ in (3.1.9) fulfills the following conditions:

$$|\mu(z)| < 1\, ,$$

(3.1.12)

and the region of absolute stability l is defined as

$$l = \{z \mid (3.1.12) \text{ holds}\}\, .$$

(3.1.13)

Definition 3.8. $R(z)$ is an approximation of order p if there exists a constant $C \neq 0$ such that

$$\exp(z) - R(z) = Cz^{p+1} + 0(z^{p+2})\, ,$$

(3.1.14)

and the approximation is consistent if it is at least of order one (i.e., $p \geq 1$).

Definition 3.9. The numerical integration scheme (3.1.1) is said to be A-stable provided the region of absolute stability specified in (3.1.13) includes the entire left half of the complex z plane.

A-stability is a very desirable property for any numerical integration algorithm, particularly if the IVP (2.1.4) were to be stiff or stiff oscillatory (i.e., an inherently stable differential system (2.1.4) in which the product of the Lipschitz constant and the interval of integration is very large).

The justification for Definitions 3.8 and 3.9 may be attributed to the existence of a similarity matrix H such that $y = Hz$ transforms the linear differential equation

$$y\prime = Ay \qquad (3.1.15)$$

into the uncoupled system

$$z\prime = \Lambda z , \qquad (3.1.16)$$

where

$$\Lambda = H^{-1}AH \qquad (3.1.17)$$

is a diagonal matrix.

It must be emphasized that practical applications of the stability of the linearized problem cannot be expected to be foolproof, although, by definition, the stability results are always foolproof.

Lambert (1987) identified the limitations and abuses of absolute stability theory, and he advocated the adoption of the one-sided Lipschitz condition (2.2.2b).

In case the Lipschitz function $V(x) \le 0$ for some $x \, e \, [a, b]$, the solutions are expected to behave contractively.

Dahlquist (1959) suggested the adoption of logarithmic norm

$$\mu(A) = \lim_{\delta \to 0} [I + \delta A - 1]/\delta , \qquad (3.1.18)$$

which accommodates negative values. Aiken (1985) identified the important properties of the logarithmic norm (pages 265-266). Chapter 8 highlights the work of Dahlquist and others on the stability and error analysis for stiff nonlinear problems.

3.2. The Euler Scheme, the Inverse Euler Scheme, and Richardson's Extrapolation

The numerical integration methods to be discussed will be expected to satisfy any desirable stability concept.

Euler proposed the simplest and the most analyzed numerical integration algorithm. It is a one-step scheme (3.1.1) whose increment function $\phi_E(x, y; h)$ is specified as

$$\phi_E(x, y; h) = f(x, y) . \qquad (3.2.1)$$

That is,

$$y_{n+1} = y_n + hf(x_n, y_n) \qquad (3.2.2)$$

for Equation (3.1.8), whose geometrical representation is given in Figure 3.2.

The Euler scheme (3.2.2) is justified by the solution to the following integral equation:

$$y(x+s) = y(x) + \int_{x}^{x+s} f(\xi, y(\xi))d\xi .$$
(3.2.3)

Assuming that $f(\xi, y(\xi))$ is constant in the interval $[x, x+s]$ and setting $x = x_n$ and $s = h$ in (3.2.3) imply (3.2.2).

Another explanation for (3.2.2) is that it is a first order Taylor's series expansion of $y(x_{n+1})$ about $x = x_n$.

The stability function $\mu_E(z)$ for Euler's scheme is given by

$$\mu_E(z) = 1 + z ,$$
(3.2.4)

and the interval of absolute stability is $l = \{z \, \varepsilon \, C : \, |\mu_E(z)| < 1\}$.

The convergence of the Euler scheme can be verified, since its application to the scalar test problem $y' = \lambda y$ results in the following first order difference equation:

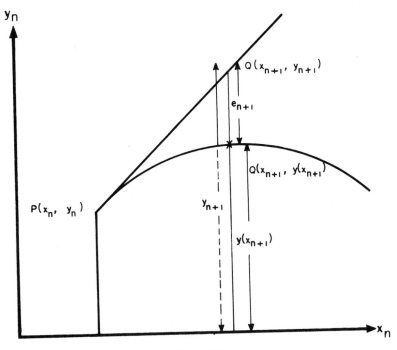

Figure 3.2

$$y_n = (1 + \lambda h)^n y_0 , \qquad (3.2.5)$$

which, in the limit, does satisfy

$$\lim_{\substack{n \to \infty \\ h \to 0}} y_n = y(x_n) . \qquad (3.2.6)$$

The analytic proof of the convergence of the Euler scheme is given in Section 3.3.

The fact that the explicit Euler scheme (3.2.2) has a small interval of absolute stability renders it unsuitable for stiff systems. A significant improvement can be attained by adopting the implicit Euler scheme

$$y_{n+1} = y_n + hf(x_{n+1}, y_{n+1}) , \qquad (3.2.7)$$

whose stability function $\mu_1(z)$ can be obtained as

$$\mu_1(z) = \frac{1}{1-z} , \qquad (3.2.8)$$

which is the $(0, 1)$ Padé approximation to e^z, with interval of absolute stability being the complex left half-plane.

The implementation of the implicit Euler is discussed in Section 8.3 and requires the solution of nonlinear systems.

An alternative to the backward Euler scheme is the inverse Euler method proposed independently by Lambert (1973a,b) and Fatunla (1982a) and given as

$$y_{n+1} = \frac{y_n^2}{y_n - hy_n'} = y_n + \frac{hy_n y_n'}{y_n - hy_n'} . \qquad (3.2.9)$$

The inverse Euler method (3.2.9) is component-applicable to systems of differential equations.

The stability property of (3.2.9) is the same as that of the implicit Euler for scalar ODEs, while it eliminates the need for the solution of nonlinear systems. A serious drawback of (3.2.9) is that it is not valid if $y_n - hy_n' \approx 0$, but we can adjust h to get around this problem.

The main drawback of both the Euler and inverse Euler schemes is that they are both of order one, which is rather low. However, this situation can be significantly improved by the application of the Richardson extrapolation to the limit of either of the two schemes. Gragg (1964, 1965) proved that, in general, one step integration schemes, when applied to the IVP (2.1.4) with a uniform meshsize over a finite interval $a \leq x \leq b$, possess an asymptotic error expansion of the form

$$y(b, h) = y(b) + \sum_{r=J}^{\infty} \gamma_r h^{\tau_r} , \quad J \geq 1 , \qquad (3.2.10)$$

with $\tau_r = \delta r$ expressible in either of the following ways:

$$\tau_r = \begin{cases} r, & \delta = 1, \\ 2r, & \delta = 2. \end{cases} \qquad (3.2.11)$$

$y(b, h)$ is the numerical estimate of the exact solution $y(b)$ obtained with meshsize h, and γ_r are constants independent of h.

The integration process is repeated over the interval $[a,b]$ with a smaller meshsize αh for $0 < \alpha < 1$. This gives the following alternative solution to (3.2.10):

$$y(b, \alpha h) = y(b) + \sum_{r=J}^{\infty} \gamma_r (\alpha h)^{\tau_r}, \quad J \geq 1. \qquad (3.2.12)$$

Multiply (3.2.12) by $(1/\alpha)^{\delta J}$ and then subtract (3.2.10) to get

$$\frac{(\frac{1}{\alpha})^{\delta J} y(b, \alpha h) - y(b, h)}{(\frac{1}{\alpha})^{\delta J} - 1} = y(b) + \sum_{r=J+1}^{\infty} \overline{\gamma}_r h^{\tau_r}, \qquad (3.2.13)$$

$$\overline{\gamma}_r = \frac{((\frac{1}{\alpha})^{\delta(r-J)} - 1)}{(\frac{1}{\alpha})^{\delta J} - 1} \gamma_r. \qquad (3.2.14)$$

For sufficiently small h, Equation (3.2.13) gives a more accurate solution than either the Euler scheme or the inverse Euler scheme, as is evident from Table 3.1. There are more advanced extrapolation schemes, namely the polynomial extrapolation scheme independently proposed by Neville (1934) and Aitken (1932), and the rational extrapolation scheme proposed by Bulirsch and Stoer (1966a,b).

Table 3.1. $y' = -y$, $y(0) = 1$, $0 \leq x \leq 1$
Euler Scheme Richardson Extrapolation $\tau_r = r$

h	$y(1,h)$	$y(1,h/2)$	$2y(1,h/2)$ $-y(1,h)$	error	error/h^2
1/2	0.25000000	0.31640625	0.38281250	0.01493305	0.05973220
1/4	0.31640625	0.34360892	0.37081158	0.00293213	0.04691410
1/8	0.34360892	0.35607415	0.36853939	0.00065994	0.04223633
1/16	0.35607415	0.36205527	0.36803639	0.00015694	0.04017639
1/32	0.36205527	0.36498642	0.36791757	0.00003812	0.03903198
1/64	0.36498642	0.36643767	0.36788893	0.00000948	0.03881836

Table 3.2. $y\prime = -y$, $y(0) = 1$, $0 \le x \le 1$
Inverse Euler Richardson Extrapolation $\tau_r = r$

h	$y(1,h)$	$y(1,h/2)$	$2y(1,h/2)$ $-y(1,h)$	error	error/h^2
1/2	0.44444448	0.40960002	0.37475556	0.00687611	0.02750444
1/4	0.40960002	0.38974431	0.36988860	0.00200915	0.03214645
1/8	0.38974431	0.37908536	0.36842641	0.00054696	0.03500557
1/16	0.37908536	0.37355384	0.36802232	0.00014287	0.03657532
1/32	0.37355384	0.37073502	0.36791620	0.00003675	0.03762817
1/64	0.37073502	0.36931172	0.36788842	0.00000897	0.03674316

It can be observed that for this particular problem, the numerical results generated with the Euler scheme approach the exact solution from below, while those of the inverse Euler approach from above.

The next section can be skipped by readers who are only interested in the application of numerical methods.

3.3. The Convergence of Euler's Scheme

The following lemma will be useful in this and subsequent sections:

Lemma 3.1. *Let $\{e_j, j = 0(1)n\}$ be a set of real numbers. If there exist finite constants R and S such that*

$$|e_{j+1}| \le R|e_j| + S , \quad j = 0(1)n-1 , \tag{3.3.1}$$

then

$$|e_j| \le \frac{R^j-1}{R-1} S + R^j|e_0| , R \ne 1 . \tag{3.3.2}$$

Proof. For $j = 0$, (3.3.2) is satisfied identically as $|e_0| \le |e_0|$.

Assume (3.3.2) holds for $j \le k$ so that

$$|e_k| \le \frac{R^k-1}{R-1} S + R^k|e_0| . \tag{3.3.3}$$

For $j = k$, (3.3.1) gives

$$|e_{k+1}| \le R|e_k| + S . \tag{3.3.4}$$

Insertion of (3.3.3) in (3.3.4) yields

$$|e_{k+1}| \le \frac{R^{k+1}-1}{R-1} S + R^{k+1}|e_0| . \tag{3.3.5}$$

Hence, (3.3.2) holds for all $j \geq 0$.   ∎

The next theorem gives the necessary conditions for the convergence of the Euler scheme.

Theorem 3.1. *Let $f(x, y)$ satisfy the hypothesis of the existence Theorem 2.1. If, in addition, $f(x, y)$ satisfies a Lipschitz condition of order one w.r.t. x, then the Euler scheme is convergent.*

Proof. Let t_{n+1} and e_{n+1} denote, respectively, the l.t.e. and the g.e. of Euler's scheme. Hence, from Definition 3.2 and Equation (3.1.3), the l.t.e. t_{n+1} for the Euler scheme is obtained as

$$y(x_{n+1}) = y(x_n) + hf(x_n, y(x_n)) + t_{n+1} . \tag{3.3.6}$$

Subtracting (3.2.2) from (3.3.6) leads to

$$e_{n+1} = e_n + h(f(x_n, y(x_n)) - f(x_n, y_n)) + t_{n+1} . \tag{3.3.7}$$

Let T be an upper bound for the l.t.e. for Euler's scheme applied to (2.1.4) in $a \leq x \leq b$; i.e.,

$$\max_{a \leq x_n \leq b} |t_n| = T . \tag{3.3.8}$$

Also denote by L and K the Lipschitz constants for $f(x, y)$ w.r.t. y and x, respectively.

By taking the norm of both sides of (3.3.7) and using the triangle inequality, the Lipschitz condition w.r.t. y and (3.3.8) lead to the following inequality:

$$|e_{n+1}| \leq (1+hL)|e_n| + T , \quad n = 0, 1,.... \tag{3.3.9}$$

Equation (3.3.9) satisfies the hypothesis of Lemma 3.1 with $R = 1 + hL$, and $S = T$; i.e.,

$$|e_n| \leq \frac{(1+hL)^n - 1}{hL} T + (1+hL)^n |e_0| . \tag{3.3.10}$$

For real z, $1 + z \leq e^z$. Therefore, $(1+hL)^n \leq e^{nhL} = e^{L(x_n-a)}$.

Since $x_n \leq b$, we have

$$e^{L(x_n-a)} \leq e^{L(b-a)} . \tag{3.3.11}$$

The application of (3.3.11) in (3.3.10) gives the following bound on the global error:

$$|e_n| \leq \frac{e^{L(b-a)} - 1}{hL} T + e^{L(b-a)} |e_0| . \tag{3.3.12}$$

A more illuminating bound for the global error can be derived by using more analytic properties of the IVP to obtain a sharper estimate for T.

Using the Mean Value Theorem in (3.3.6) yields

$$t_{n+1} = h\left(f\left(x_n+\theta h,\, y\left(x_n+\theta h\right)\right) - f\left(x_n,\, y\left(x_n\right)\right)\right), \quad 0 \le \theta \le 1.$$

We now add and subtract the quantity $hf\left(x_n+\theta h,\, y\left(x_n\right)\right)$ to the righthand side of the last equation, and, in addition, take the norm of both sides of the equation—the Lipschitz conditions on $f\left(x,\, y\right)$—in both variables to get

$$|t_{n+1}| \le h\,|f\left(x_n+\theta h,\, y\left(x_n+\theta h\right)\right) - f\left(x_n+\theta h,\, y\left(x_n\right)\right)|,$$

$$+ h\,|f\left(x_n+\theta h,\, y\left(x_n\right)\right) - f\left(x_n,\, y\left(x_n\right)\right)|,$$

$$\le hL\,|y\left(x_n+\theta h\right) - y\left(x_n\right)| + Kh^2\theta,$$

$$= Lh^2\theta\,|y\prime(\xi)| + Kh^2\theta,\quad x_n \le \xi \le x_{n+1}.$$

Let Y denote the upper bound for $y\prime(x)$ in $a \le x \le b$; i.e.,

$$Y = \max_{a \le x \le b}\, |y\prime(x)|. \tag{3.3.13}$$

By adopting this in the previous expression for l.t.e., we obtain

$$|t_{n+1}| \le h^2\theta(LY+K)$$

$$\le h^2(LY+K)$$

$$= T.$$

By substituting $T = h^2(LY+K)$ in (3.3.12), the g.e. then satisfies the relation

$$|e_n| \le h\,\frac{e^{L(b-a)}-1}{L}\,(LY+K) + e^{L(b-a)}\,|e_0|, \tag{3.3.14}$$

which, in the limit, satisfies

$$\lim_{\substack{h\to 0 \\ |e_0|\to 0}} e_n = 0. \tag{3.3.15}$$

From (3.1.5) and (3.1.6), (3.3.15) implies

$$\lim_{\substack{h\to 0 \\ |e_0|\to 0}} y_n = y\left(x_n\right),$$

thus establishing the convergence of Euler's scheme. ∎

Henrici (1962, p. 71) established the following theorem concerning the convergence of the general one-step method (3.1.1):

Theorem 3.2. *Let the increment function* $\Phi(x,\, y;\, h)$ *be continuous in each of its arguments for* $(x,\, y)\, \varepsilon\, R^{m+1}$ *and* $0 < h \le h_0$, *and, in addition, let* Φ *satisfy a*

Lipschitz condition of order one with respect to y. Then, (3.1.1) is convergent if and only if it is consistent.

3.4. The Trapezoidal Scheme

Another important and very useful numerical integration formula is the trapezoidal scheme, whose increment function ϕ_{TR} is given by

$$\phi_{TR}(x_n, y_n; h) = \frac{1}{2}(f(x_n, y_n) + f(x_{n+1}, y_{n+1})) . \tag{3.4.1}$$

This implies

$$y_{n+1} = y_n + \frac{h}{2}(f(x_n, y_n) + f(x_{n+1}, y_{n+1})) , \tag{3.4.2}$$

which is implicit since the unknown quantity y_{n+1} occurs implicitly on the right hand side of (3.4.2).

Equation (3.4.2) can be derived from the following representation of the integration formula:

$$\alpha_1 y_{n+1} + \alpha_0 y_n = h(\beta_1 f_{n+1} + \beta_0 f_n) , \quad \alpha_1 = 1 , \tag{3.4.3}$$

where α_0, β_0 and β_1 are real, undetermined coefficients and $f_{n+j} \equiv f(x_{n+j}, y_{n+j})$, $j = 0, 1$.

The numerical values of these undetermined coefficients can be obtained either

(a) by Taylor expansion of the terms on both sides of (3.4.3) about $x = x_n$ and then by equating the terms in powers of h, or

(b) by constraining (3.4.3) to be exact whenever the theoretical solution $y(x)$ is a polynomial of degree ≤ 2.

Both procedures lead to the following set of equations in α_0, β_0, β_1:

$$1 + \alpha_0 = 0 ,$$

$$\beta_0 + \beta_1 = 1 , \tag{3.4.4}$$

$$2\beta_1 = 1 ,$$

whose solution is $\alpha_0 = -1$, $\beta_0 = \beta_1 = \frac{1}{2}$.

As (3.4.2) is implicit, the solution y_{n+1} has to be generated iteratively.

An initial estimate $y_{n+1}^{[0]}$ can be given by any of the following explicit one-step formulas:

$$y_{n+1}^{[0]} = \begin{cases} y_n, \\ y_n + hf_n, \\ \dfrac{y_n^2}{y_n - hf_n}, y_n \neq hf_n, \end{cases} \tag{3.4.5}$$

and this can then be modified recursively as follows:

$$y_{n+1}^{[s+1]} = y_n + \frac{h}{2}\left[f(x_n, y_n) + f(x_{n+1}, y_{n+1}^{[s]})\right], \quad s = 0, 1, 2,..., \tag{3.4.6}$$

possibly until certain convergence criteria are met—for instance, until $|y_{n+1}^{[s+1]} - y_{n+1}^{[s]}| \leq \varepsilon$, where ε is some allowable tolerance and $y_{n+1} = y_{n+1}^{[s+1]}$. It is also possible that a fixed number of iterations M is prescribed; i.e., $y_{n+1} = y_{n+1}^{[M]}$.

The numerical integration formula (3.4.6) was implemented adopting the predictors specified by (3.4.5), i.e.,

(i)   (3.4.5a) and (3.4.6) with $M = 1$ (EXPTR-1),

(ii)   (3.4.5b) and (3.4.6) with $M = 1$ (EXPTR-2),

(iii) (3.4.5c) and (3.4.6) with $M = 1$ (EXPTR-3),

(iv)   (3.4.5a) and (3.4.8) with Newton Iteration (IMPTR-1).

Although the explicit trapezoidal schemes EXPTR-1, EXPTR-2, and EXPTR-3 all demand less computational effort than the implicit trapezoidal scheme IMPTR-1, the latter has better stability properties, as is evident from Table 3.3.

Lambert (1973a) and Shampine (1978) noted that the restriction on $|\lambda h|$ for convergence of (3.4.6) is entirely analogous to the stability constraint. That is,

Table 3.3.

Method	Number of Iterations	Stability Function	Interval of Absolute Stability
	M	$\mu(z)$	
EXPTR-1	1	$1 + z$	$(-2, 0)$
EXPTR-2	1	$1 + z + 1/2z^2$	$(-2, 0)$
EXPTR-3	1	$\dfrac{1 - 1/2z^2}{1 - z}$	$(-3.236, 0)$
IMPTR-1	>1	$\dfrac{1 + 1/2z}{1 - 1/2z}$	$(-\infty, 0)$

$|\lambda h| < 2$ for simple iteration to converge in general. The stability intervals quoted in Table 3.3 have not taken this into consideration.

The superiority of the implicit trapezoidal scheme IMPTR can be better appreciated by considering the application of all four integration schemes to the following initial value problem:

$$y\prime = -1000(y-x^3) + 3x^2 \, , \quad y(0) = 0 \, , \qquad (3.4.7)$$

$0 \le x \le 1$ adopting uniform meshsizes $\{h = 2^{-r}, r = 2(2)12\}$. The theoretical solution of IVP (3.4.7) is $y(x) = x^3$.

The details of the numerical results are given in Table 3.4.

The basic explanation for the outcome of Table 3.4 (why the explicit trapezoidal schemes fail to give accurate results for meshsizes $h > 2^{-10}$) is that the corresponding z falls outside the interval of absolute stability. EXPTR-3 yields a fairly accurate result for $h = 2^{-8}$, simply because the stability consideration compels it to adopt a meshsize $h \le 0.003236$, and, hence, the adoption of $h = 2^{-8} = 0.003906$ throws the corresponding z outside the interval of absolute stability. Stability consideration demands the adoption of $h \le 2 \times 10^{-3}$ for EXPTR-1 and EXPTR-2, while IMPTR-1 is insulated from any stability constraint on h. Hence, all the numerical estimates for $h = 2^{-r}$, $r = 2(2)12$ gave reasonable results for the implicit trapezoidal rule.

For large stepsize, nonlinear equations have to be solved, and a brief discussion on this can be found in Section 8.3.

However, the adoption of the Newton-Raphson scheme (8.3.7) and (8.3.10) in the implicit trapezoidal scheme (3.4.6) leads to the following numerical integration procedure (called IMPTR-1):

$$y_{n+1}^{[s+1]} = y_{n+1}^{[s]} - \left[I - \frac{1}{2}h \, \frac{\partial f}{\partial y} (x_{n+1}, y_{n+1}^{[s]}) \right]^{-1}$$

$$(3.4.8)$$

Table 3.4. $y\prime = -1000(y-x^3) + 3x^2$, $y(0) = 0$, $0 \le x \le 1$
Theoretical Solution is $y(x) = x^3$; Numerical Estimate to $y(1)$

h	EXPTR-1	EXPTR-2	EXPTR-3	IMPTR-1
2^{-2}	∞	∞	∞	1.00000193738
2^{-4}	∞	∞	∞	1.00000125189
2^{-6}	∞	∞	∞	1.00000012207
2^{-8}	∞	∞	1.33181734745	1.00000000763
2^{-10}	1.00146334559	1.00000279188	0.99999436540	1.00000000048
2^{-12}	1.00036556851	1.00000010176	0.99999979633	1.00000000003

$$\cdot \{y_{n+1}^{[s]} - y_n - \frac{1}{2}h\left[f(x_n, y_n) + f(x_{n+1}, y_{n+1}^{[s]})\right]\},$$

which converges faster than formula (3.4.6), but at a cost of solving linear systems. This is due to the fact that (3.4.8) has second order convergence, while (3.4.6) converges linearly.

Although implicit methods require more computational effort, they possess important stability properties that make them useful for certain category of problems.

Dahlquist (1963) established the following Barrier Theorem:

Theorem 3.3.

(i) An explicit k-step method cannot be A-stable.

(ii) The order p of an A-stable, linear multistep method cannot exceed 2. The smallest error constant $C^* = 1/12$ is obtained for the trapezoidal rule (3.4.2).

This is not to say that the scheme is faultless. For instance, its stability function $\mu(z)$ is given in Table 3.3 as

$$\mu(z) = \frac{1+\frac{1}{2}z}{1-\frac{1}{2}z}, \qquad (3.4.9)$$

which is the $(1, 1)$ Padé approximation to e^z with

$$\lim_{z \to -\infty} \mu(z) = -1,$$

which suggests that the trapezoidal scheme oscillates at infinity.

Liniger and Willoughby (1970) adopted a variation to the trapezoidal scheme with Equation (3.4.2) being replaced by IMPTR-2, which is specified as

$$y_{n+1} = y_n + h\left[\theta f(x_n, y_n) + (1-\theta)f(x_{n+1}, y_{n+1})\right], \qquad (3.4.10)$$

whose stability function is expressed as

$$\mu_\theta(z) = \frac{1+\theta z}{1-(1-\theta)z}. \qquad (3.4.11)$$

IMPTR-2 is A-stable provided $\theta \leq 1/2$. However, the authors suggested that θ be chosen so as to ensure that $\mu_\theta(z)$ is equal to the stiff component e^z of the IVP. This leads to the following choice of θ:

$$\theta = -\frac{e^z}{1-e^z} - \frac{1}{z}. \qquad (3.4.12)$$

Gourlay (1970) revealed another defect of the conventional trapezoidal scheme on its application to the test problem

$$y' = -\lambda(x)y , \ \lambda(x) \ge 0 ,$$
(3.4.13)

which yields the following recurrence relation:

$$y_{n+1} = \left[\frac{1 - \frac{1}{2}h\lambda(x_n)}{1 + \frac{1}{2}h\lambda(x_{n+1})} \right] y_n .$$
(3.4.14)

Stability demands that

$$h[\lambda(x_n) - \lambda(x_{n+1})] \le 4 ,$$
(3.4.15)

which imposes a severe constraint on the meshsize h, particularly when $\lambda(x)$ is large and rapidly decaying. He analyzed the implicit midpoint rule

$$y_{n+1} = y_n + hf(x_{n+1/2}, \frac{1}{2}(y_n + y_{n+1})) ,$$
(3.4.16)

which is tagged IMPTR-3. The integration formula (3.4.16) is insulated from a possibly severe meshsize constraint imposed on IMPTR-1 by Inequality (3.4.15).

Consider the IVP (3.4.13) with $\lambda(x)$ specified as

$$\lambda(x) = 100(100-x) , \quad 0 \le x \le 1 ,$$
(3.4.17)

whose theoretical solution is given by

$$y(x) = \exp\left[-100x(100 - \frac{x}{2}) \right] .$$

Gourlay (1970) observed that the conventional trapezoidal scheme is defective on this problem, in the sense that the Constraint (3.4.15) compels the scheme to adopt a meshsize $h \le 0.2$, while IMPTR-3 specified by Equation (3.4.16) could adopt meshsize $h \ge 0.2$.

Liniger (1958), Nevanlinna and Liniger (1978, 1979), and Dahlquist (1975) introduced a similar concept to Gourlay's (1970). Liniger specified the contractivity criteria for a stable IVP: that two numerical solutions y_n and z_n should satisfy

$$y_n - z_n \le y_{n-1} - z_{n-1}$$
(3.4.18)

for some suitable norm $\cdot$.

If the conventional trapezoidal rule (3.4.6) were applied to the IVP (2.1.4) with $f(x, y)$ being nonlinear, contractivity requires that

$$\left[I - \frac{1}{2}hJ_{n+1}\right]^{-1} \left[I + \frac{1}{2}hJ_n\right] \leq 1 . \tag{3.4.19}$$

Apart from the fact that the Constraint (3.4.19) is difficult to fulfill, particularly if an eigenvalue of the Jacobian varies rapidly in the interval $[x_n, x_{n+1}]$, the constraint is liable to be violated. A prototype of the one-leg methods proposed by Dahlquist is given by (3.4.16), and the contractivity condition requires that

$$\left[I - \frac{1}{2}hJ_{n+1/2}\right]^{-1} \left[I + \frac{1}{2}hJ_{n+1/2}\right] \leq 1 . \tag{3.4.20}$$

Condition (3.4.20) is readily fulfilled in some norm, provided that (2.1.4) is stable; i.e., $\text{Re}(\lambda(J_n)) \leq 0$, and J_n is diagonalizable.

4. RUNGE-KUTTA PROCESSES

4.1. General Theory of Runge-Kutta Processes

The Taylor expansion method of order p for the numerical solution of the initial value problem (2.1.4) is given by

$$y_{n+1} = y_n + h\Phi_T(x_n, y_n; h) , \qquad (4.1.1)$$

with

$$\Phi_T(x_n, y_n; h) = \sum_{r=0}^{p-1} \frac{h^r}{(r+1)!} f^{(r)}(x_n, y_n) , \qquad (4.1.2a)$$

where $f^{(r)}(x_n, y_n)$, $r = 0(1)p-1$ denotes the r-th derivative of $f(x, y(x))$; i.e.,

$$f^{(r)}(x, y(x)) = y^{(r+1)}(x) = (\frac{\partial}{\partial x} + f\frac{\partial}{\partial y})^r f(x, y(x)) . \qquad (4.1.2b)$$

The local truncation error t_{n+1} for Method (4.1.1) can be obtained as

$$t_{n+1} = \frac{h^{p+1}}{(p+1)!} f^{(p)}(x_n, y_n) + O(h^{p+2}) . \qquad (4.1.3)$$

The need to generate analytically first and higher order derivatives of $f(x, y)$ often makes the Taylor expansion method impractical. However, symbolic generation of higher order derivatives of $f(x, y)$ can be obtained for an extensive range of problems.

In the case of linear initial value problems of the form

$$y\prime = A(x)y + d(x) \tag{4.1.4}$$

(where A is an $m \times m$ matrix and y, $d(x)$ are m-vectors), the first and higher order derivatives of $f(x, y)$ can be obtained recursively as follows:

$$f^{(k)} = \sum_{r=1}^{k} \binom{k}{r} A^{(r)}(x) y^{(k-r)} + d^{(k)}(x) \tag{4.1.5}$$

for $i = 1(1)m$.

In the example of the differential equation

$$y\prime = x^2 + y^2 \ ,$$

the higher order derivatives can be generated recursively and compactly as

$$y\prime\prime = 2x + yy\prime + y\prime y \ ,$$

$$y\prime\prime\prime = 2 + yy\prime\prime + 2y\prime y\prime + y\prime\prime y \ ,$$

$$y^{(4)} = yy\prime\prime\prime + 3y\prime y\prime\prime + 3y\prime\prime y\prime + y\prime\prime\prime y \ ,$$

$$\vdots$$

$$y^{(k+1)} = \sum_{j=0}^{k} \binom{k}{j} y^{(j)} y^{(k-j)} \ , \ k \geq 3 \ .$$

Halin (1983) discussed efficient methods of expressing higher derivatives in terms of lower ones in very general cases, including systems of ODEs with piecewise continuous righthand sides.

A snag with the Taylor expansion method, and, indeed with any method, is that the order can be restricted if, for some $r < p$, the r-th derivative is undefined at a particular point within the interval of integration. For the initial value problem

$$y\prime = (x-a)^{3/2} y \ , \ y(a) = 1 \ ,$$

the Taylor expansion method of order $p > 2$ cannot be formed at $x = a$.

An s-stage Runge-Kutta (R-K) process can be viewed as an extension of the Taylor expansion scheme whereby the evaluation of the first and higher order derivatives of $f(x, y)$ is replaced by s function evaluations within every interval of integration $[x_n, x_{n+1}]$.

The R-K scheme is basically a substitution method of the form

$$y_{n+1} = y_n + h\Phi_{RK}(x_n, y_n; h) \ , \tag{4.1.6a}$$

with the increment function Φ_{RK} given as a weighted mean of the slopes at specific points

$$\{z_r \mid z_r = x_n + c_r h \ , \ 0 \le c_r \le 1\} :$$

$$\Phi_{RK} = \sum_{j=1}^{s} b_j Y_j \ , \ s \ge 1 \ , \tag{4.1.6b}$$

with the constraint

$$\sum_{j=1}^{s} b_j = 1 \ . \tag{4.1.6c}$$

Here, s is the number of stages of the process, and Condition (4.1.6c) ensures that the integration scheme is consistent (i.e., at least of order one).

The slopes $Y_r, r = 1(1)s$ are defined by

$$Y_r = f\,(z_r, \ y_n + h \sum_{j=1}^{s} a_{rj} Y_j) \ , \tag{4.1.6d}$$

$$z_r = x_n + c_r h \ , \ r = 1(1)s \ , \tag{4.1.6e}$$

and

$$c_r = \sum_{j=1}^{s} a_{rj} \ , \ r = 1(1)s \ . \tag{4.1.6f}$$

Runge-Kutta methods can be classified into three categories (see Equation (4.4.1) and Table 4.1).

The numerical values of the unknown coefficients $\{b_j, c_r, a_{rj}\}$ are normally obtained from a set of nonlinear equations, derived as follows:

Table 4.1. Coefficients of R-K Methods

Type	Status of Coefficient	Number of Coefficients
Explicit	$a_{rj}=0$, $j \ge r$ i.e., A is lower triangular with zero diagonal elements	$s(s+1)/2$
Semi-Implicit	$a_{rj} = 0$, $j > r$ A is lower triangular with non-zero diagonal elements	$s(s+3)/2$
Implicit	$a_{rj} \ne 0$ for at least one $j > r$; i.e., A is not a lower triangular matrix	$s(s+1)$

Step 1

Obtain the Taylor series expansion of Y_r (defined by (4.1.6d)) about the point (x_n, y_n) for $r = 1(1)s$.

Step 2

Insert these expansions and (4.1.6f) in (4.1.6b).

Step 3

Compare the coefficients in powers of h for both the increment function Φ_{RK} of the Runge-Kutta method given by (4.1.6b) and the increment function Φ_T for the Taylor expansion method specified by (4.1.2).

The totality of the unknown coefficients $\{b_j, c_r, a_{rj}; r, j = 1(1)s\}$ normally exceeds the number of equations, so some can be chosen so as to attain some desired goal. Some of these goals are

(i)   to minimize a bound of the local truncation error (cf. Ralston 1962),

(ii) to maximize the attainable order of the scheme (King (1966) achieved this for the differential systems $y\prime = f(x)$),

(iii) to optimize the interval of absolute stability (Lawson 1966, 1967b), and

(iv) to reduce storage requirements (Gill 1951, Conte and Reeves 1956, Blum 1962, and Fyfe 1966).

Butcher (1965) established the following relationship between the stage s and the order p of the explicit Runge-Kutta processes:

$$p(s) = s, \ 1 \leq s \leq 4,$$

$$p(s) \leq s-1, \ 5 \leq s \leq 7, \text{ and}$$

$$p(s) \leq s-2, \ s \geq 8.$$

Butcher (1963, 1976b), Hairer (1978), Wanner et al. (1978), and Hairer and Wanner (1981) also used the concept of rooted trees (in graph-theoretic sense) to establish the order conditions for all classes of R-K processes:

$$C(p): \ \sum_{j=1}^{s} a_{ij} c_j^{k-1} = \frac{c_i^k}{k}, \ i = 1(1)s, \ k \leq p,$$

$$D(p): \ \sum_{i=1}^{s} b_i c_i^{k-1} a_{ij} = \frac{b_j(1-c_j^k)}{k}, \ j = 1(1)s, \ k \leq p,$$

$$B(p): \ \sum_{i=1}^{s} b_i c_i^{k-1} = \frac{1}{k}, \ k = 1(1)p.$$

Lemma 4.1. $C(\eta)$, $D(\xi)$, $B(p)$ *for* $p \leq \xi + \eta + 1$, $p \leq 2\eta + 2$ *implies that the R-K process (4.1.6) is of order p.*

There is a rapid growth of number of conditions with p.

Error Estimation for R-K Processes

Lotkin (1951) proposed the following error bound for the partial derivatives of $f(x, y)$, which could be useful in obtaining an upper bound for the local truncation errors of the Runge-Kutta schemes:

Suppose the differential system (2.1.4) satisfies

$$|f(x, y)| \leq M , \tag{4.1.7a}$$

and

$$|\frac{\partial f}{\partial y}| \leq L \tag{4.1.7b}$$

for $a \leq x \leq b$ and $y < \infty$.

Lotkin assumes the inequality

$$|\frac{\partial^{i+j} f(x, y)}{\partial x^i \partial y^j}| < L^{i+j} M^{1-j} \tag{4.1.8}$$

in order to get a tidy bound for various methods. The Lotkin error bound is merely a theoretical tool for comparing methods.

For all the one-step methods (like R-K methods), the conceptually-simplest definition of local truncation error (l.t.e.) is that it is the error committed in the most recent integration step on a single integration step.

Denote by $y(x; a, \alpha)$ the solution to the initial value problem

$$y\prime = f(x, y) , y(a) = \alpha . \tag{4.1.9}$$

The local truncation error (l.t.e.) for the one-step method

$$y_{n+1} = y_n + h\Phi(x_n, y_n; h) \tag{4.1.10}$$

is given by

$$\text{l.t.e.} = y_{n+1} - y(x_{n+1}; x_n, y_n) . \tag{4.1.11}$$

For h sufficiently small, a reasonable estimate of this local truncation error can be obtained as follows:

1. Compute y_{n+1} from (x_n, y_n) using meshsize h; denote this by $y(x_{n+1}, h)$ using (4.1.10).

2. Compute $\hat{y}_{n+1}$ from (x_n, y_n) in two steps using meshsize $h/2$, and denote this by $y(x_{n+1}, h/2)$:

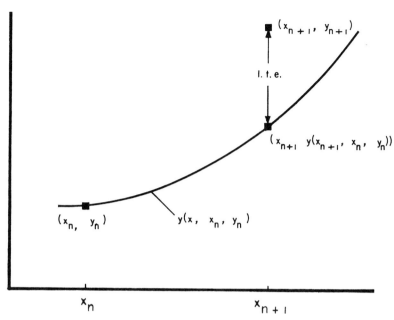

Figure 4.1: Local Truncation Error

$$y_{n+1/2} = y_n + \frac{h}{2}\Phi(x_n, y_n; \frac{h}{2}) \,,$$

$$\hat{y}_{n+1} = y(x_{n+1}, h/2) = y_{n+1/2} + h/2 \,\Phi(x_{n+1/2}, y_{n+1/2}; h/2) \,.$$

3. Compute an estimate of the local truncation error as

$$\text{l.t.e.} = \frac{y(x_{n+1}, \frac{h}{2}) - y(x_{n+1}, h)}{2^p - 1}$$

$$= y_{n+1} - y(x_{n+1}; x_n, y_n) + O(h^{p+2}) \,. \tag{4.1.12}$$

Equation (4.1.12) is an estimate of the local error for the two steps that give $y(x_{n+1,}\, h/2)$.

It must be noted that

$$y_{n+1} = y(x_{n+1}; x_n, y_n) + C(x_{n+1}) \cdot h^p + O(h^{p+1})$$

and

$$\hat{y}_{n+1} = y(x_{n+1}; x_n, y_n) + C(x_{n+1}) \cdot (\frac{h}{2})^p + O(h^{p+1}) \,.$$

Then

$$\text{l.t.e.} = \hat{y}_{n+1} - y(x_{n+1}; x_n, y_n)$$

$$\approx C(x_{n+1}) \cdot (\frac{h}{2})^p$$

$$= C(x_n)(\frac{h}{2})^p + C\prime(x_n) \cdot 2 \cdot (\frac{h}{2})^{p+1} + \ldots$$

$$= O(h^{p+1}),$$

since $C(x_n) = 0$.

If two solutions $y(x_n, h)$ and $y(x_n, h/2)$ (for $n = 1, 2, 3, \ldots$) are computed in parallel, then the global error (g.e.)

$$\text{g.e.} = C(x_n) \cdot h^p$$

can be estimated by

$$\text{g.e.} \approx \frac{y(x_n, \frac{h}{2}) - y(x_n, h)}{2^p - 1}. \tag{4.1.13}$$

This is a quantity that estimates the error in $y(x_n, h/2)$ committed after the point where $C(x_n) = 0$, i.e.,

(1) from the starting point if two independent solutions are computed, and

(2) from x_{n-1} if both solutions start at $(x_{n-1}, y(x_{n-1}, h/2))$.

Local error estimates can be efficiently computed using the embedding approach originally proposed by Sarafyan (1965), Shintani (1966), Fehlberg (1968, 1969, 1970), and England (1969). An s-stage method of order p is embedded in $(p+1)$-th order scheme with stage $\geq s+1$ such that the difference in the numerical solutions constitutes the local error estimate for the lower order scheme. Unfortunately, for IVPs ($y\prime = f(x)$) that reduce to evaluation of quadratures, Fehlberg's method of order greater than five gives error estimates which are identically zero, and, thus, are unreliable for such problems. Verner (1978, 1979) developed methods of arbitrarily high orders of accuracy whose error estimators, based on the embedding approach, overcome this difficulty. Both Fehlberg and Verner based their work on explicit R-K processes, with some derivative evaluations of the two methods being identical. Shampine and Watts (1976c, 1977b, 1979a) developed automatic codes RKF45 and its offspring DEGRK based on Fehlberg (1970) (4,5) pair, while Hull, Enright, and Jackson (1980) developed the code DVERK based on R-K formulas of orders five and six of Verner. DVERK is now included in the IMSL. A version of the RKF45 code appears in Forsythe et al. (1977), and Gladwell (1979) developed the code D02PAF based on the Fehlberg (1970) (4,5) pair for the NAG software.

The Fehlberg (1970) (p,q) pair approach is as follows:

$$y_{n+1} = y_n + h_n \sum_{i=1}^{s} b_i Y_i \text{ , order } p \text{ ,} \tag{4.1.14}$$

$$\hat{y}_{n+1} = y_n + h_n \sum_{i=1}^{s} \hat{b}_i Y_i \text{ , order } q = p+1 \text{ ,} \tag{4.1.15}$$

where (4.1.14) is an R-K process of error order p, and (4.1.15) is of error order $q = p+1$, using the same internal approximations Y_i, $i = 1(1)s$. The local error estimator is

$$\text{l.t.e.} \approx \hat{y}_{n+1} - y_{n+1} \text{ .} \tag{4.1.16}$$

In case $\hat{y}_{n+1}$ is adopted as the approximate solution, then we speak of "local extrapolation." In the alternative,

$$\hat{y}_{n+1} = y_{n+1} + h_n E_{n+1} \text{ ,}$$

where E_{n+1} is the local error estimate per unit step.

Many popular R-K codes—D02PAF (Gladwell 1979), RKF45 (Shampine and Watts 1979a), and DVERK (Hull et al. 1976)—adopt error per step criterion and local extrapolation.

Enright and Hull (1976) felt that an embedded pair of formulas of orders four and five due to Fehlberg is much more efficient than the classical four-stage, fourth-order R-K formula with doubling or Richardson extrapolation. Deuflhard (1983) disagreed with this view and argued that it is hard to construct even one good formula of high order, harder to construct a pair. Kaps et al. (1985) asserted that the two procedures are very much the same, and that, for other kinds of one-step methods, stability is crucial, and it may be difficult to construct an embedding pair with both having good stability.

Shampine (1984d) used the fourth-order explicit R-K methods to obtain an appraisal of doubling. He provided an elementary proof of the doubling error estimator and used this to establish the existence of a higher order result at the middle of the integration step. An explicit R-K formula with error estimated by doubling is simply a general way to construct an embedding pair.

The Fehlberg (1970) (4,5) pair given by (4.4.14) has shown up very well in extensive numerical tests. Apart from the efficiency of the Dormand and Prince (1980) pair specified by (4.4.15), Shampine (1986) established that the DP (4,5) pair readily accommodates the "interpolation" earlier applied to the Fehlberg pair in Horn (1983) and discussed in Shampine (1985c). Shampine and Baca (1984b) developed an experimental code XRK based on the Dormand and Prince (1981) (7,8) pair (4.4.16). XRK was observed to be at least as efficient as the extrapolation code DIFEX1 (Deuflhard 1985).

4.2. The Explicit Two-Stage Process

The previous discussion can be better appreciated by illustrating the derivation of the two-stage explicit R-K process.

The explicit two-stage R-K process is given as

$$y_{n+1} = y_n + h\Phi_{RK2}(x_n, y_n; h) , \tag{4.2.1a}$$

whose increment function Φ_{RK2} is given by

$$\Phi_{RK2} = b_1 Y_1 + b_2 Y_2 , \tag{4.2.1b}$$

with the constraint

$$b_1 + b_2 = 1 . \tag{4.2.1c}$$

The slopes Y_1 and Y_2 are specified as

$$Y_1 = f(x_n, y_n) , \tag{4.2.1d}$$

and

$$Y_2 = f(x_n + c_2 h, y_n + ha_{21} Y_1) , \tag{4.2.1e}$$

with the constraint

$$c_2 = a_{21} . \tag{4.2.1f}$$

The Taylor series expansion of Y_2 about the point (x_n, y_n) in the solution space yields

$$Y_2 = f + h(c_2 f_x + a_{21} f f_y)$$

$$\tag{4.2.2}$$

$$+ \frac{1}{2} h^2 (c_2^2 f_{xx} + 2c_2 a_{21} f f_{xy} + a_{21}^2 f^2 f_{yy}) + O(h^3) ,$$

with all the terms evaluated at (x_n, y_n).

We substitute (4.2.1d) and (4.2.2) in (4.2.1b) to get

$$\Phi_{RK2} = (b_1 + b_2)f + h(b_2 c_2 f_x + b_2 a_{21} f f_y)$$

$$\tag{4.2.3}$$

$$+ \frac{1}{2} h^2 (b_2 c_2^2 f_{xx} + 2b_2 c_2 a_{21} f f_{xy} + b_2 a_{21}^2 f^2 f_{yy}) + O(h^3) .$$

The equivalent increment function Φ_{T2} for the Taylor expansion method can be obtained from (4.1.2) as

$$\Phi_{T2} = f + \frac{1}{2} h(f_x + f f_y) + \frac{1}{6} h^2 (f_{xx} + 2f f_{xy}$$

$$\tag{4.2.4}$$

$$+ f^2 f_{yy} + f_x f_y + ff_y^2) + O(h^3) .$$

All the terms of (4.2.3) and (4.2.4) are evaluated at the point (x_n, y_n).

By requiring as many terms as possible to coincide in the two increment functions, Φ_{RK2} for the two-stage R-K process and Φ_{T2} for the Taylor expansion method of order two, the following set of equations emerges:

$$b_1 + b_2 = 1 ,$$

$$b_2 c_2 = \frac{1}{2} . \tag{4.2.5}$$

Set (4.2.5) consists of two equations in three unknowns, leaving us with one degree of freedom. We have to adopt (4.2.1f); otherwise, there is no solutions to the equations.

The two-stage explicit R-K process is a second order scheme with the principal error function obtained from (4.2.3) and (4.2.4) as

$$\psi_2(x_n, y_n; h) = \frac{1}{6} [3b_2 c_2^2 f_{xx} + 6b_2 c_2 a_{21} ff_{xy}$$

$$\tag{4.2.6}$$

$$+ 3b_2 a_{21}^2 f^2 f_{yy} - f_{xx} - 2ff_{xy} - f^2 f_{yy} - f_x f_y - ff_y^2](x_n, y_n) .$$

By adopting (4.2.1f) and (4.2.5) in (4.2.6) and nominating $\alpha = c_2 = a_{21}$ as the free parameter, the error function ψ_2 for the explicit two-stage R-K process (4.2.1) is obtained as

$$\psi_2(x_n, y_n; h) = \frac{1}{6} [(\frac{3}{2} f_{xx} + 3ff_{xy} + \frac{3}{2} f^2 f_{yy}) \alpha$$

$$\tag{4.2.7}$$

$$- (f_{xx} + 2ff_{xy} + f^2 f_{yy}) - (f_x f_y + ff_y^2)]_{(x_n, y_n)} .$$

The following bounds for the partial derivatives of $f(x, y)$ can be obtained with a bound in the form (4.1.8) used by Lotkin:

$$|f_{xx}| < ML^2 ,$$

$$|f_{xy}| < L^2 ,$$

and

$$|f_{yy}| < M^{-1} L^2 . \tag{4.2.8}$$

Inserting the inequalities (4.1.7), (4.1.8), and (4.2.8) into (4.2.7) leads to the following bound on the error function ψ_2:

$$|\psi_2(x_n, y_n; h)| < [|\alpha - \frac{2}{3}| + \frac{1}{3}]ML^2 ,\qquad (4.2.9)$$

which attains a minimum $1/3ML^2$ when the free parameter is $\alpha = 2/3$. Hence, the coefficients of an optimal two-stage R-K process are obtained as

$$\alpha = c_2 = a_{21} = \frac{2}{3} , \ b_1 = \frac{1}{4}, \ b_2 = \frac{3}{4} .$$

The resulting method is

(i) $$y_{n+1} = y_n + \frac{h}{4}(Y_1 + 3Y_2) ,$$

$$Y_1 = f(x_n, y_n) , \qquad (4.2.10)$$

$$Y_2 = f(x_n + \frac{2}{3}h , y_n + \frac{2}{3}hY_1) .$$

Method (4.2.10) is called the "optimal" two-stage Runge-Kutta process or Heun two-stage scheme.

Other commonly used two-stage R-K schemes include

(ii)   the explicit midpoint rule whereby

$$\alpha = c_2 = a_{21} = \frac{1}{2} , b_1 = 0 , b_2 = 1 ;$$

i.e.,

$$y_{n+1} = y_n + hf(x_n + \frac{1}{2}h , y_n + \frac{1}{2}hf(x_n, y_n)) , \qquad (4.2.11a)$$

often called improved Euler or Heun method.

Other methods which aspire to this name include

$$y_{n+1} = y_{n-1} + 2hf(x_n, y_n) , \ \text{(explicit)} \qquad (4.2.11b)$$

and

$$y_{n+1} = y_n + hf(x_{n+1/2}, \frac{1}{2}(y_n + y_{n+1})) , \ \text{(semi–implicit)} . \qquad (4.2.11c)$$

(iii) $$\alpha = c_2 = a_{21} = 1; b_1 = b_2 = \frac{1}{2} ;$$

i.e.,

$$y_{n+1} = y_n + \frac{1}{2}h[f(x_n, y_n) + f(x_n + h, y_n + hf(x_n, y_n))] . \qquad (4.2.12)$$

It is now clear that there exist infinitely many two-stage Runge-Kutta processes of order two, depending on choice of the free parameter α.

Example. Determine the appropriate meshsize so as to integrate the initial value problem

$$y\prime = -10(y-1)^2 \; , \; y(0) = 2 \; , \; 0 \le x \le 1 \; , \tag{4.2.13}$$

with the optimal two-stage R-K scheme (4.2.10) with an allowable error tolerance $\varepsilon = 10^{-4}$.

The theoretical solution to Problem (4.2.13) is

$$y(x) = 1 + \frac{1}{1+10x} \; . \tag{4.2.14}$$

We readily establish the following bounds: $M = 10$, $L = 20$, and

$$|\psi_2(x_n, y_n, h)| \le \frac{1}{3} ML^2$$

$$= \frac{1}{3} \cdot 10 \cdot 20^2 \; .$$

The principal error term is hence bounded as

$$|\psi_2(x_n, y_n; h)| h^3 \le \frac{1}{3} \cdot 10 \cdot 20^2 \cdot h^3 \le 10^{-4} \; ,$$

which confines the meshsize h to values

$$h \le 0.004217 \; .$$

In this particular example, the error depends on the solution in the sense that ψ_2 depends on f, and so one cannot, in general, predetermine the meshsize.

The accuracy can be improved with the application of Richardson extrapolation (cf. Section 3.2) to the Runge-Kutta schemes. This yields the

Table 4.2. Numerical Results of Example
$y\prime = -10(y-1)^2$, $y(0) = 2$, $0 \le x \le 1$; Uniform Meshsize $h = 0.0025$

x_n	$10^7 \times \psi(x_n, y_n; h)$	$10^4 \times (y(x_n, h) - y(x_n))$	$10^4 \times$ exact error
0.1	5.77870	0.52901	0.53575
0.2	1.12834	0.31519	0.31651
0.3	0.35494	0.19956	0.19998
0.4	0.14488	0.13622	0.13639
0.5	0.06971	0.09852	0.09860
0.6	0.03756	0.07443	0.07448
0.7	0.02199	0.05816	0.05819
0.8	0.01372	0.04668	0.04669
0.9	0.00899	0.03828	0.03829
1.0	0.00614	0.03195	0.03196

improved solution

$$\overline{y}_{n+1} = \frac{2^p y (x_{n+1}, \frac{h}{2}) - y (x_{n+1}, h)}{2^p - 1},$$

where p is the order of the R-K process, and $y(x_{n+1}, h)$ is the numerical estimate (generated with meshsize h) to the exact solution $y(x_{n+1})$.

4.3. Convergence and Stability of the Two-Stage Explicit R-K Scheme (4.2.1)

The properties of the increment function Φ of a one-step scheme are, in general, very crucial to its stability and convergence characteristics.

Lemma 4.2. *If the initial value problem (2.1.4) satisfies the hypothesis of the Existence and Uniqueness Theorem (Theorem 2.1), then the increment function Φ_{RK2} specified by (4.2.1b) satisfies a Lipschitz condition of order one w.r.t. to the dependent variable y.*

Proof. Denote by L a Lipschitz constant for $f(x, y)$ w.r.t. y and set

$$Y_1(y_n) = f(x_n, y_n). \tag{4.3.1}$$

Then

$$|Y_1(y_n) - Y_1(z_n)| = |f(x_n, y_n) - f(x_n, z_n)| \tag{4.3.2}$$

$$\leq L |y_n - z_n|.$$

Similarly,

$$|Y_2(y_n) - Y_2(z_n)| = |f(x_n + c_2 h, y_n + h a_{21} Y_1(y_n))$$

$$- f(x_n + c_2 h, z_n + h a_{21} Y_1(z_n))| \tag{4.3.3}$$

$$\leq L\{|y_n - z_n| + h |a_{21}| |Y_1(y_n) - Y_1(z_n)|\}$$

$$\leq L |y_n - z_n| \{1 + hL |a_{21}|\}.$$

We also have

$$|\Phi_{RK2}(x_n, y_n; h) - \Phi_{RK2}(x_n, z_n; h)|$$

$$= |b_1 Y_1(y_n) + b_2 Y_2(y_n) - b_1 Y_1(z_n) - b_2 Y_2(z_n)|$$

$$\leq |b_1| |Y_1(y_n) - Y_1(z_n)| + |b_2| |Y_2(y_n) - Y_2(z_n)|.$$

By inserting relations (4.3.2) and (4.3.3) in the last inequality, we obtain

$$| \Phi_{RK2}(x_n, y_n; h) - \Phi_{RK2}(x_n, z_n; h)| \leq L^* | y_n - z_n | , \qquad (4.3.4)$$

with the Lipschitz constant L^* given as

$$L^* = L\{|b_1| + |b_2|(1+hL|a_{211}|)\} . \qquad (4.3.5)$$

∎

Lemma 4.2 holds for all classes of R-K methods regardless of the stage number.

The following theorem guarantees the stability of the Runge-Kutta processes.

Theorem 4.1. *If the initial value problem (2.1.4) fulfills the hypothesis of the Existence Theorem (Theorem 2.1), then Method (4.2.1) is stable in the sense of Definition (3.6).*

Proof. Let $\{y_n\}$ and $\{z_n\}$ be two sets of solutions generated recursively by the two-stage R-K scheme (4.2.1) with the initial conditions $y(a) = y_0$, $z(a) = z_0$, $|y_0 - z_0| = \delta_0$.

Let

$$\delta_n = y_n - z_n , \ n \geq 0 , \qquad (4.3.6)$$

and

$$y_{n+1} = y_n + h\Phi_{RK2}(x_n, y_n; h) , \qquad (4.3.7)$$

$$z_{n+1} = z_n + h\Phi_{RK2}(x_n, z_n; h) . \qquad (4.3.8)$$

Taking the norm of the difference between (4.3.7) and (4.3.8) and invoking the triangle inequality implies

$$|\delta_{n+1}| \leq (1+hL^*)|\delta_n| , \ n \geq 0 . \qquad (4.3.9)$$

The application of Lemma 3.1 to the Inequality (4.3.9) with $R = 1+hL^*$, $S = 0$ gives

$$|\delta_n| \leq K |\delta_0| , \qquad (4.3.10)$$

where

$$K = e^{L^*}(b-a) < \alpha , \qquad (4.3.11)$$

which, from Definition 3.6, implies the stability of Method (4.2.1). ∎

Theorem 4.1 can be extended to all classes of R-K schemes.

The application of Method (4.2.1) to the scalar test problem $y' = \lambda y$ readily yields the first order difference equation

$$y_{n+1} = \mu_2(z)y_n , \ z = \lambda h , \qquad (4.3.12)$$

where the stability function $\mu_2(z)$ is

Table 4.3. Stability Intervals for Explicit R-K Methods		
Stages s	$\mu_s(z)$	Interval of Absolute Stability
1	$1 + z$	$(-2, 0)$
2	$1 + z + 1/2z^2$	$(-2, 0)$
3	$1 + z + 1/2z^2 + 1/6z^3$	$(-2.51, 0)$
4	$1 + z + 1/2z^2 + 1/6z^3 + 1/24z^4$	$(-2.78, 0)$

$$\mu_2(z) = 1 + z + \frac{1}{2}z^2 , \tag{4.3.13}$$

which fulfills

$$|\mu_2(z)| < 1 ,$$

provided $-2 \leq z \leq 0$. This implies that the interval of absolute stability for all Methods (4.2.1) is $l = (-2, 0)$.

Table 4.3 gives the interval of absolute stability for the explicit R-K processes with stages $s \leq 4$ and order $p = s$.

Theorem 3.2 gives the necessary and sufficient condition for the convergence of the Runge-Kutta processes in general.

4.4. Matrix Representation of the R-K Processes

An s-stage Runge-Kutta process given by (4.1.6) can be described by the matrix

$$\begin{pmatrix} A & c \\ b^{\mathrm{T}} & 0 \end{pmatrix} = \begin{pmatrix} a_{11} & a_{12} & \cdots & a_{1s} & a_{1s+1} \\ a_{21} & & & & \vdots \\ \vdots & & & & \vdots \\ a_{s1} & & \cdots & a_{ss} & a_{ss+1} \\ a_{s+11} & & \cdots & a_{s+1s} & 0 \end{pmatrix} , \tag{4.4.1a}$$

whose coefficients are specified as follows:

$$
\begin{aligned}
a_{rs+1} &= c_r, & r &= 1(1)s \\
a_{s+1j} &= b_j, & j &= 1(1)s \\
a_{s+1s+1} &= 0, & & \\
a_{ij} &= a_{ij}, & i,j &= 1(1)s
\end{aligned}
$$

(4.4.1b)

Some explicit R-K schemes are represented as follows:

One-stage Method

Euler

$$
\begin{bmatrix}
0 & 0 \\
1 & 0
\end{bmatrix}.
$$

(4.4.2)

Two-stage Methods

Midpoint Rule

$$
\begin{bmatrix}
0 & 0 & 0 \\
\dfrac{1}{2} & 0 & \dfrac{1}{2} \\
0 & 1 & 0
\end{bmatrix},
$$

(4.4.3)

Optimal or Heun

$$
\begin{bmatrix}
0 & 0 & 0 \\
\dfrac{2}{3} & 0 & \dfrac{2}{3} \\
\dfrac{1}{4} & \dfrac{3}{4} & 0
\end{bmatrix},
$$

(4.4.4)

Explicit Trapezoidal

$$\begin{bmatrix} 0 & 0 & 0 \\ 1 & 0 & 1 \\ \dfrac{1}{2} & \dfrac{1}{2} & 0 \end{bmatrix} . \tag{4.4.5}$$

Three-stage Processes

Optimal or Ralston Scheme

$$\begin{bmatrix} 0 & 0 & 0 & 0 \\ \dfrac{1}{2} & 0 & 0 & \dfrac{1}{2} \\ 0 & \dfrac{3}{4} & 0 & \dfrac{3}{4} \\ \dfrac{2}{9} & \dfrac{1}{3} & \dfrac{4}{9} & 0 \end{bmatrix} , \tag{4.4.6}$$

Heun Scheme

$$\begin{bmatrix} 0 & 0 & 0 & 0 \\ \dfrac{1}{3} & 0 & 0 & \dfrac{1}{3} \\ 0 & \dfrac{2}{3} & 0 & \dfrac{2}{3} \\ \dfrac{1}{4} & 0 & \dfrac{3}{4} & 0 \end{bmatrix} , \tag{4.4.7}$$

Classical

$$\begin{bmatrix} 0 & 0 & 0 & 0 \\ \dfrac{1}{2} & 0 & 0 & \dfrac{1}{2} \\ -1 & 2 & 0 & 1 \\ \dfrac{1}{6} & \dfrac{4}{6} & \dfrac{1}{6} & 0 \end{bmatrix} , \tag{4.4.8}$$

Nystrom

$$\begin{bmatrix} 0 & 0 & 0 & 0 \\ \dfrac{2}{3} & 0 & 0 & \dfrac{2}{3} \\ 0 & \dfrac{2}{3} & 0 & \dfrac{2}{3} \\ \dfrac{2}{8} & \dfrac{3}{8} & \dfrac{3}{8} & 0 \end{bmatrix} . \qquad (4.4.9)$$

Four-stage Schemes

Classical

$$\begin{bmatrix} 0 & 0 & 0 & 0 & 0 \\ \dfrac{1}{2} & 0 & 0 & 0 & \dfrac{1}{2} \\ 0 & \dfrac{1}{2} & 0 & 0 & \dfrac{1}{2} \\ 0 & 0 & 1 & 0 & 1 \\ \dfrac{1}{6} & \dfrac{2}{6} & \dfrac{2}{6} & \dfrac{1}{6} & 0 \end{bmatrix} , \qquad (4.4.10)$$

Kutta

$$\begin{bmatrix} 0 & 0 & 0 & 0 & 0 \\ \dfrac{1}{3} & 0 & 0 & 0 & \dfrac{1}{3} \\ -\dfrac{1}{3} & 1 & 0 & 0 & \dfrac{2}{3} \\ 1 & -1 & 1 & 0 & 1 \\ \dfrac{1}{8} & \dfrac{3}{8} & \dfrac{3}{8} & \dfrac{1}{8} & 0 \end{bmatrix} , \qquad (4.4.11)$$

Gill

$$\begin{bmatrix}
0 & 0 & 0 & 0 & 0 \\
\dfrac{1}{2} & 0 & 0 & 0 & \dfrac{1}{2} \\
\dfrac{\sqrt{2}-1}{2} & \dfrac{2-\sqrt{2}}{2} & 0 & 0 & \dfrac{1}{2} \\
0 & \dfrac{-\sqrt{2}}{2} & \dfrac{2+\sqrt{2}}{2} & 0 & 1 \\
\dfrac{1}{6} & \dfrac{2-\sqrt{2}}{6} & \dfrac{2+\sqrt{2}}{6} & \dfrac{1}{6} & 0
\end{bmatrix} , \qquad (4.4.12)$$

Kutta Merson Scheme

$$\begin{bmatrix}
0 & 0 & 0 & 0 & 0 & 0 \\
\dfrac{1}{3} & 0 & 0 & 0 & 0 & \dfrac{1}{3} \\
\dfrac{1}{6} & \dfrac{1}{6} & 0 & 0 & 0 & \dfrac{1}{3} \\
\dfrac{1}{8} & 0 & \dfrac{3}{3} & 0 & 0 & \dfrac{1}{2} \\
\dfrac{1}{2} & 0 & -\dfrac{3}{2} & 2 & 0 & 1 \\
\dfrac{1}{6} & 0 & 0 & \dfrac{4}{6} & \dfrac{1}{6} & 0
\end{bmatrix} . \qquad (4.4.13)$$

Matrix representation of R-K processes provides a compact way of describing such schemes.

Coefficients of Fehlberg's (1970) (4,5) Pair (4.4.14)

				$-\dfrac{845}{4104}$		0	0
0	0	0	0	0	0	0	0
$\dfrac{1}{4}$	0	0	0	0	0	0	$\dfrac{1}{4}$
$\dfrac{3}{32}$	$\dfrac{9}{32}$	0	0	0	0	0	$\dfrac{3}{8}$
$\dfrac{1932}{2197}$	$-\dfrac{7200}{2197}$	$\dfrac{7296}{2197}$	0	0	0	0	$\dfrac{12}{13}$
$\dfrac{439}{216}$	-8	$\dfrac{3680}{513}$	$-\dfrac{845}{4104}$	0	0	0	1
$-\dfrac{8}{27}$	2	$-\dfrac{3544}{2565}$	$\dfrac{1859}{4104}$	$-\dfrac{11}{40}$	0	0	$\dfrac{1}{2}$

| b_i | $\dfrac{25}{216}$ | 0 | $\dfrac{1408}{2565}$ | $\dfrac{2197}{4104}$ | $-\dfrac{1}{5}$ | 0 | 0 |
| $\hat{b}_i$ | $\dfrac{16}{135}$ | 0 | $\dfrac{6656}{12825}$ | $\dfrac{28561}{56430}$ | $-\dfrac{9}{50}$ | $\dfrac{2}{55}$ | 0 |

Coefficients of Dormand and Prince (1980) Seven-Stage Scheme (4.4.15)

R-K5 (4) 7M with Interpolation (cf. Shampine 1986)

Results at midpoint is computed according to $y_{n+1/2} = y_n + \frac{1}{2}h \sum_{i=1}^{7} b_i^* Y_i$

	0	0	0	0
	$\dfrac{1}{5}$	0	0	0
	$\dfrac{3}{40}$	$\dfrac{9}{40}$	0	0
	$\dfrac{44}{45}$	$-\dfrac{56}{15}$	$\dfrac{32}{9}$	0
	$\dfrac{19372}{6561}$	$-\dfrac{25360}{2187}$	$\dfrac{64448}{6561}$	$-\dfrac{212}{729}$
	$\dfrac{9017}{3168}$	$-\dfrac{355}{33}$	$\dfrac{46732}{5247}$	$\dfrac{49}{176}$
	$\dfrac{35}{384}$	0	$\dfrac{500}{1113}$	$\dfrac{125}{192}$
$\hat{b}_i$ P = 5 at x_{n+1}	$\dfrac{35}{384}$	0	$\dfrac{500}{1113}$	$\dfrac{125}{192}$
$\hat{b}_i$ P = 4 at x_{n+1}	$\dfrac{1951}{21600}$	0	$\dfrac{22642}{50085}$	$\dfrac{451}{720}$
$\hat{b}_i$ P = 4 at k_{n+1}	$\dfrac{6025192743}{30085553152}$	0	$\dfrac{51252292925}{65400821598}$	$-\dfrac{2691868925}{45128329728}$

Coefficients of Dormand and Prince (1980) (4.4.15) continued

				c_i
	0	0	0	0
	0	0	0	$\frac{1}{5}$
	0	0	0	$\frac{3}{10}$
	0	0	0	$\frac{4}{5}$
	0	0	0	$\frac{8}{9}$
	$-\dfrac{5103}{18656}$	0	0	1
	$-\dfrac{2187}{6784}$	$\dfrac{11}{84}$	0	1
P = 5 at x_{n+1} $\hat{b}_i$	$-\dfrac{2187}{6784}$	$\dfrac{11}{84}$	0	0
P = 4 at x_{n+1} $\hat{b}_i$	$-\dfrac{12231}{42400}$	$\dfrac{649}{6300}$	$\dfrac{1}{60}$	0
P = 4 at k_{n+1} $\hat{b}_i$	$\dfrac{187940372067}{1594534317056}$	$-\dfrac{1776094331}{19743664256}$	$\dfrac{11237099}{235043384}$	0

Coefficients of Prince and Dormand (1981)
R-K8 (7) 13M.Code XRK (Shampine 1984d) (4.4.16)

	0	0	0	0	0	0	0
	$\dfrac{1}{18}$	0	0	0	0	0	
	$\dfrac{1}{48}$	$\dfrac{1}{16}$	0	0	0		
	$\dfrac{1}{32}$	0	$\dfrac{3}{32}$	0	0		
	$\dfrac{5}{16}$	0	$\dfrac{-75}{64}$	$\dfrac{75}{64}$			
	$\dfrac{3}{80}$	0	0	$\dfrac{3}{16}$	$\dfrac{3}{20}$	0	0
	$\dfrac{29443841}{614563906}$	0	0	$\dfrac{77736538}{692538347}$	$\dfrac{-28693883}{1125000000}$	$\dfrac{23124283}{1800000000}$	0
	$\dfrac{16016141}{946692911}$	0	0	$\dfrac{61564180}{158712637}$	$\dfrac{22789713}{633445777}$	$\dfrac{545815736}{2771057229}$	$\dfrac{-180193667}{1043307555}$
	$\dfrac{39632708}{573591083}$	0	0	$\dfrac{-433036366}{683701615}$	$\dfrac{-421739975}{2616792301}$	$\dfrac{100302831}{723423059}$	$\dfrac{790204164}{839813087}$
	$\dfrac{246121993}{134047787}$	0	0	$\dfrac{-37695042795}{15268766246}$	$\dfrac{-309121744}{1061227803}$	$\dfrac{-12992083}{490766935}$	$\dfrac{6005943493}{2108947869}$
	$\dfrac{-1028468189}{846180014}$	0	0	$\dfrac{8478235783}{508512852}$	$\dfrac{1311729495}{1432422823}$	$\dfrac{-10304129875}{1701304382}$	$\dfrac{-48777925059}{3047939560}$
	$\dfrac{185892177}{718116043}$	0	0	$\dfrac{-3185094517}{667107341}$	$\dfrac{-477755414}{1098053517}$	$\dfrac{-703635378}{230739211}$	$\dfrac{5731566787}{1027545527}$
	$\dfrac{403863854}{491063109}$	0	0	$\dfrac{-5068492393}{434740067}$	$\dfrac{-411421997}{543043805}$	$\dfrac{652783627}{914296604}$	$\dfrac{11173962825}{925320556}$
b_i	$\dfrac{13451932}{455176623}$	0	0	0	0	$\dfrac{-808719846}{976000145}$	$\dfrac{1757004468}{5645159321}$
$\hat{b}_i$	$\dfrac{14005451}{335480064}$	0	0	0	0	$\dfrac{-59238493}{1068277825}$	$\dfrac{181606767}{758867731}$

Coefficients of Prince (1981) (4.4.16) continued

0	0	0	0	0	0	0
0	0	0	0	0	0	$\frac{1}{18}$
0	0	0	0	0	0	$\frac{1}{12}$
0	0	0	0	0	0	$\frac{1}{8}$
0	0	0	0	0	0	$\frac{5}{16}$
0	0	0	0	0	0	$\frac{3}{8}$
0	0	0	0	0	0	$\frac{39}{400}$
0	0	0	0	0	0	$\frac{93}{200}$
$\frac{800635310}{3783071287}$	0	0	0	0	0	$\frac{5490023248}{9719169821}$
$\frac{393006217}{1396673457}$	$\frac{123872331}{1001029789}$	0	0	0	0	$\frac{13}{20}$
$\frac{15336726248}{1032824649}$	$\frac{-45442868181}{3398467696}$	$\frac{3065993473}{597172653}$	0	0	0	$\frac{1201146811}{1299019798}$
$\frac{5232866602}{850066563}$	$\frac{-4093664535}{808688257}$	$\frac{3962137247}{1805957418}$	$\frac{65686358}{487910083}$	0	0	1
$\frac{-13158990841}{6184727034}$	$\frac{3936647629}{1978049680}$	$\frac{-160528059}{685178525}$	$\frac{248638103}{1413531060}$	0	0	0

b_i	$\frac{656045339}{2658591186}$	$\frac{-3867574721}{1518517206}$	$\frac{465885868}{322736530}$	$\frac{53011238}{667516719}$	$\frac{2}{45}$	0	1
$\hat{b}_i$	$\frac{561292985}{797845732}$	$\frac{-1041891430}{1371343529}$	$\frac{760417231}{1151165299}$	$\frac{118820643}{751138087}$	$\frac{-528747749}{2220607170}$	0	0

4.5. Error Estimation and Stepsize Selection in R-K Processes

Ralston (1962) provided an error bound (in the spirit of Lotkin (4.1.8)) for the classical fourth-order Runge-Kutta scheme (4.4.10):

$$| \psi(x_n, y_n; h) | \leq \frac{73}{720} ML^4 . \qquad (4.5.1)$$

In case of the scalar initial value problem:

$$y' = y , \; y(0) = 1 , \; 0 \leq x \leq 1 , \; \varepsilon = 10^{-4} ,$$

$M = \exp(1)$ and $L = 1$. The leading error term satisfies the relation

$$| \psi(x_n, y_n; h) h^5 | < 10^{-4} ,$$

provided $h \leq 0.206$.

These error bounds are good only for theoretical comparison of methods and are not meant for choosing stepsizes in practical problems.

An alternative approach to meshsize selection is to apply the Runge-Kutta scheme to the same scalar initial value problem to obtain

$$Y_1 = f(x_n, y_n) = y_n ,$$

$$Y_2 = f(x_n + \frac{1}{2} h, \; y_n + \frac{1}{2} h Y_1)$$

$$= (1 + \frac{1}{2} h) y_n ,$$

$$Y_3 = f(x_n + \frac{1}{2} h, \; y_n + \frac{1}{2} h Y_2)$$

$$= (1 + \frac{1}{2} h + \frac{1}{4} h^2) y_n ,$$

$$Y_4 = f(x_n + h, \; y_n + h Y_3)$$

$$= (1 + h + \frac{1}{2} h^2 + \frac{1}{4} h^3) y_n .$$

These slopes can be inserted into the integration formula

$$y_{n+1} = y_n + \frac{h}{6} (Y_1 + 2Y_2 + 2Y_3 + Y_4)$$

to get

$$y_{n+1} = (1 + h + \frac{1}{2} h^2 + \frac{1}{6} h^3 + \frac{1}{24} h^4) y_n . \qquad (4.5.2)$$

The local truncation error for the classical fourth order R-K scheme is

$$y(x_{n+1}) = (1+h+\frac{1}{2}h^2+\frac{1}{6}h^3+\frac{1}{24}h^4)y(x_n) - t_{n+1} . \quad (4.5.3)$$

Subtracting (4.5.3) from (4.5.2) and taking norms yields the following relationship:

$$|e_{n+1}| \le (1+h+\frac{1}{2}h^2+\frac{1}{6}h^3+\frac{1}{24}h^4)|e_n| + |t_{n+1}| . \quad (4.5.4)$$

The application of the Mean Value Theorem to (4.5.3) gives the local truncation error as

$$t_{n+1} = -\frac{h^5}{120}y(\xi) , \; x_n \le \xi \le x_{n+1} .$$

(This result holds for vectors also.) The Relation (4.5.4) is thence reduced to

$$|e_{n+1}| \le (1+h+\frac{1}{2}h^2+\frac{1}{6}h^3+\frac{1}{24}h^4)|e_n| + \frac{1}{120}h^5 e . \quad (4.5.5)$$

The application of Lemma 2.1 (with $e_0 = 0$) to (4.5.5) results in the following error bound:

$$|e_N| \le \frac{h^5}{120} \frac{(1+h+\frac{1}{2}h^2+\frac{1}{6}h^3+\frac{1}{24}h^4)^N - 1}{(h+\frac{1}{2}h^2+\frac{1}{6}h^3+\frac{1}{24}h^4)} e$$

$$\le \frac{h^5}{120} \frac{e^{Nh} - 1}{h} e$$

$$= \frac{h^4}{120} e(e-1) .$$

The requirement that $|e_N| \le 10^{-4}$ is fulfilled provided $h \le 0.225$,

Numerical experiments were carried out with uniform meshsizes 0.20 and 0.225, and Richardson's error estimator was used. Details are given in Table 4.4.

The results presented by Shampine (1972a, 1973) and Shampine and Gordon (1975) indicate that the use of local extrapolation (whereby the local error estimate for a step is added to the numerical approximation) can significantly improve the performance of a method. This is at the expense of forfeiting the error estimate and the loss of A-stability property. Dahlquist (1963) points out that A-stability for the trapezoidal rule is lost by local Richardson extrapolation.

The Kutta-Merson scheme (4.4.13) is a five-stage explicit R-K scheme of order four whose extra function evaluation provides an error estimator

$$t_{n+1} = \frac{h}{30}(-2Y_1+9Y_3-8Y_4+Y_5) , \qquad (4.5.6)$$

which Scraton (1964) discovered to be reliable only for linear differential systems with constant coefficients.

Shampine and Watt (1971) advocated the use of a local error estimator proposed by Ceschino and Kuntzmann (1966), which is based on quadrature formulas. With $y_{n+1}^{(0)} = y_n$, a meshsize $h_n/3$ is adopted to generate $y_{n+1}^{(1)}$, $y_{n+1}^{(2)}$, and $y_{n+1}^{(3)}$ at $x_n + (ih_n)/3$, $i = 1(1)3$, whose corresponding derivatives are

$$f_{n+1}^{(i)} = f (x_n + \frac{(ih_n)}{3} , y_{n+1}^{(i)}) . \qquad (4.5.7)$$

The local error per unit step from x_n to x_{n+1} was given as

$$E_{n+1}(h_n) = \frac{1}{20}[f_{n+1}^{(3)}+9f_{n+1}^{(2)}+9f_{n+1}^{(1)}+f_{n+1}^{(0)}]$$

$$(4.5.8)$$

$$- \frac{1}{20h_n}[11y_{n+1}^{(3)}+27y_{n+1}^{(2)}-27y_{n+1}^{(1)}-11y_{n+1}^{(0)}] .$$

Equation (4.5.8) is based upon a four-point sixth-order quadrature formula

Table 4.4. Errors in Classical Four-Stage R-K Scheme (4.4.10)
$y' = y$, $y(0) = 1$, $0 \leq x \leq 1$

x_n	$10^4 \times e(x_n)$	$10^4 \times \psi(x_n, y_n; h)h^5$
	$h = 0.20$	
0.2	-0.027351	-0.028610
0.4	-0.036466	-0.036955
0.6	-0.050480	-0.052452
0.8	-0.064894	-0.066757
1.0	-0.079504	-0.081062
	$h = 0.225$	
0.225	-0.049301	-0.050068
0.450	-0.066934	-0.069141
0.675	-0.089089	-0.087023
0.900	-0.117221	-0.119209
1.00	-0.036584	-0.038147

which satisfies

$$E_{n+1}(h_n) = t_{n+1}(h_n) + O(h_n^{p+1}) \tag{4.5.9}$$

for any R-K scheme of order $p \leq 5$. Ceschino-Kuntzmann is very expensive.

4.6. Implicit and Semi-Implicit R-K Processes

The disadvantages of the explicit R-K methods include

(a) The relatively large number of function evaluations at every integration step. A p-th order explicit R-K scheme requires at least p function evaluations per integration step, whereas a corresponding Adams method will require one function evaluation per step, but usually with more overhead costs and smaller stepsizes.

(b) The relatively small interval of absolute stability (as evident from Table 4.3) renders them unsuitable for stiff initial value problems.

In the recent past, there was a dearth of efficient and reliable error estimators for the Runge-Kutta methods. Such estimators are available in other methods like linear multistep methods. This problem is being adequately handled by the embedding approach (cf. Section 4.1).

The handicap can be adequately taken care of with the adoption of implicit or semi-implicit Runge-Kutta or Rosenbrock schemes, at the expense of solving a set of ms or m nonlinear/linear equations.

Cash (1983a) proposed a class of variable order explicit block R-K formulas for nonstiff IVPs, and Cash (1986) proposed an efficient fourth-order (explicit) block R-K formula that permits interpolation. Cash (1983b) extended this idea to develop efficient block DIRK formulas for stiff systems.

As for the implicit R-K processes, the following iteration scheme has to be solved for the slopes Y_r:

$$Y_r = f\left(x_n + c_r h, \; y_n + h \sum_{j=1}^{s} a_{rj} Y_j\right) \tag{4.6.1}$$

for $r = 1(1)s$.

Butcher (1964) established that (4.6.1) is convergent provided the meshsize h is chosen so as to ensure that the relation

$$hL\left(\max_{r} \sum_{j=1}^{s} |a_{rj}|\right) < 1 \tag{4.6.2}$$

(L being the Lipschitz constant for $f(x, y)$ w.r.t. y) holds. This particular scheme is impractical because of the Restriction (4.6.2), which excludes even mild stiffness.

By adopting identical arguments, as in the case of the explicit two-stage R-K schemes, the implicit two-stage R-K formulas can be derived with the coefficients determined from the following set of four nonlinear equations in six unknowns $c_1, c_2, b_1, b_2, a_{21}, a_{22}$:

$$b_1 + b_2 = 1 , \qquad (4.6.3a)$$

$$b_1 c_1 + b_2 c_2 = \frac{1}{2} , \qquad (4.6.3b)$$

$$b_1 c_1^2 + b_2 c_2^2 = \frac{1}{3} , \qquad (4.6.3c)$$

$$b_1 [a_{11} c_1 + c_2(c_1 - a_{11})] + b_2 [c_1(c_2 - a_{22}) + a_{22} c_2] = \frac{1}{6} . \qquad (4.6.3d)$$

Equations (4.6.3a-d) provide two free parameters, which are chosen to attain some desirable goal.

From (4.6.3a,b,c) the following relationship between c_1 and c_2 emerges:

$$c_2 = \frac{-\dfrac{1}{2} c_1 + \dfrac{1}{3}}{-c_1 + \dfrac{1}{2}} . \qquad (4.6.4)$$

With $c_1 = 0$ we readily obtain the following two-stage implicit R-K scheme of order three:

$$\begin{bmatrix} -1 & 1 & 0 \\ \dfrac{2}{3} & 0 & \dfrac{2}{3} \\ \dfrac{1}{4} & \dfrac{3}{4} & 0 \end{bmatrix} . \qquad (4.6.5)$$

In general, an s-stage implicit R-K scheme has $s(s+1)$ distinct coefficients (see Table 4.1).

Some popular implicit R-K schemes include:

Backward Euler, $s = 1$, $p = 1$,

$$\begin{bmatrix} 1 & 1 \\ 1 & 0 \end{bmatrix} . \qquad (4.6.6)$$

Trapezoidal Scheme, $s = 2$, $p = 2$,

$$\begin{bmatrix} 1 & 0 & 1 \\ 0 & 1 & 1 \\ \dfrac{1}{2} & \dfrac{1}{2} & 0 \end{bmatrix} . \qquad (4.6.7)$$

Hammer and Hollingsworth, $s = 2$, $p = 4$,

$$
\begin{bmatrix}
\dfrac{1}{4} & \dfrac{1}{4} - \dfrac{\sqrt{3}}{6} & \dfrac{1}{2} - \dfrac{\sqrt{3}}{6} \\[2ex]
\dfrac{1}{4} + \dfrac{\sqrt{3}}{6} & \dfrac{1}{4} & \dfrac{1}{2} + \dfrac{\sqrt{3}}{6} \\[2ex]
\dfrac{1}{2} & \dfrac{1}{2} & 0
\end{bmatrix} .
\tag{4.6.8}
$$

Butcher (1964),

$$
\begin{bmatrix}
\dfrac{5}{36} & \dfrac{2}{9} - \dfrac{\sqrt{15}}{15} & \dfrac{5}{36} - \dfrac{\sqrt{15}}{30} & \dfrac{1}{2} - \dfrac{\sqrt{15}}{10} \\[2ex]
\dfrac{5}{36} + \dfrac{\sqrt{15}}{24} & \dfrac{2}{9} & \dfrac{5}{36} - \dfrac{\sqrt{15}}{14} & \dfrac{1}{2} \\[2ex]
\dfrac{5}{36} + \dfrac{\sqrt{15}}{30} & \dfrac{2}{9} + \dfrac{\sqrt{15}}{15} & \dfrac{5}{36} & \dfrac{1}{2} + \dfrac{\sqrt{15}}{10} \\[2ex]
\dfrac{5}{18} & \dfrac{4}{9} & \dfrac{5}{18} & 0
\end{bmatrix} .
\tag{4.6.9}
$$

Butcher (1964) established the existence of implicit R-K schemes of order $p = 2s$, $s \geq 1$.

The A-stability of the implicit R-K schemes can be tested by the same procedure adopted for explicit R-K methods.

In fact, the stability function for the s-stage implicit R-K scheme is a rational function $u_1(z)$ given as

$$
u_1(z) = 1 + zb^T(I - zA)^{-1}e ,
\tag{4.6.10}
$$

where $z = \lambda h$, and e is the s-vector

$$
e = (1, 1, ..., 1)^T .
$$

Any implicit R-K process whose stability function $u_1(z)$ is a diagonal (or one of the first sub-diagonals) Padé table is necessarily A-stable (cf. Ehle 1968, Chipman 1971, and Axelsson 1969).

The main snag with the implicit R-K scheme is the enormous amount of computational effort required to solve the set of ms nonlinear equations (4.6.1) for the slope Y_r. Rosenbrock (1963) and Butcher (1964) showed that the degree of implicitness and nonlinearity can be reduced considerably by ensuring that the matrix A in (4.4.1) is lower triangular. The set of nonlinear equations in (4.6.1) in these semi-implicit schemes reduces to

$$Y_r = f\left(x_n + c_r h,\ y_n + h \sum_{j=1}^{r} a_{rj} Y_j\right) \tag{4.6.11}$$

for $r = 1(1)s$.

This makes it possible to solve the r-th equation in (4.6.11) for only one unknown vector Y_r.

In the case of the two-stage semi-implicit R-K process, (4.6.3) reduces to the following four equations in five unknowns $b_1, b_2, c_1, c_2, a_{22}$:

$$b_1 + b_2 = 1, \tag{4.6.12a}$$

$$b_1 c_1 + b_2 c_2 = \frac{1}{2}, \tag{4.6.12b}$$

$$b_1 c_1^2 + b_2 c_2^2 = \frac{1}{3}, \tag{4.6.12c}$$

$$b_1 c_1^2 + b_2 c_1 (c_2 - a_{22}) + b_2 c_2 a_{22} = \frac{1}{6}. \tag{4.6.12d}$$

In arriving at (4.6.12d), we have used the facts that $c_1 = a_{11}$ and $a_{21} + a_{22} = c_2$.

Norsett (1974b) proposed a two-stage semi-implicit scheme based on (4.6.12), with an additional constraint that $a_{11} = a_{22}$. The resultant scheme, due to Norsett (1974b), is

$$\begin{bmatrix} \dfrac{3+\sqrt{3}}{6} & 0 & \dfrac{3+\sqrt{3}}{6} \\[2ex] -\dfrac{\sqrt{3}}{3} & \dfrac{3+\sqrt{3}}{6} & \dfrac{3-\sqrt{3}}{6} \\[2ex] \dfrac{1}{2} & \dfrac{1}{2} & 0 \end{bmatrix}, \tag{4.6.13}$$

whose stability function can be obtained as

$$u_S(z) = \frac{1 - \dfrac{\sqrt{3}}{3} z - \dfrac{1+\sqrt{3}}{6} z^2}{(1 - \dfrac{3+\sqrt{3}}{6} z)^2}. \tag{4.6.14}$$

Other existing semi-implicit R-K schemes include

Rosenbrock (1963)
$$\begin{bmatrix} 1+\dfrac{\sqrt{6}}{6} & 0 & 1+\dfrac{\sqrt{6}}{6} \\[2mm] 0 & 1-\dfrac{\sqrt{6}}{6} & 1-\dfrac{\sqrt{6}}{6} \\[2mm] -0.41315432 & 1.41315432 & 0 \end{bmatrix}, \, p = 3, \tag{4.6.15}$$

and

Butcher (1964)
$$\begin{bmatrix} 1 & 0 & 0 & 1 \\[1mm] \dfrac{1}{4} & \dfrac{1}{4} & 0 & \dfrac{1}{2} \\[1mm] 0 & 0 & 1 & 1 \\[1mm] \dfrac{1}{6} & \dfrac{4}{6} & \dfrac{1}{6} & 0 \end{bmatrix}, \, p = 4 . \tag{4.6.16}$$

Example. Using meshsize $\{1/8, \, 1/16, \, 1/32, \, 1/64\}$, compare the numerical solution generated with

(a) the two-stage implicit R-K process (4.6.8) proposed by Hammer and Hollingsworth (1955), and

(b) the three-stage semi-implicit scheme (4.6.16) proposed by Butcher (1964).

Other A-stable semi-implicit R-K processes were proposed by Calahan (1967), Allen and Pottle (1966), Haines (1968), Cash (1976), Bui and Poon (1981), Norsett and Thomsen (1984, 1986a,b), and Houbak et al. (1985).

Method NT-I, Norsett and Thomsen (1984, 1986b), B-stable.

Of particular interest is the L-stable (cf. Definition 8.8) mono-implicit R-K scheme of Cash (1975b). He established a one-one correspondence between the explicit and semi-implicit R-K processes.

SDIRK

	$\dfrac{5}{6}$			$\dfrac{5}{6}$
	$-\dfrac{61}{108}$	$\dfrac{5}{6}$		$\dfrac{29}{108}$
	$-\dfrac{23}{183}$	$-\dfrac{33}{61}$	$\dfrac{5}{6}$	$\dfrac{1}{6}$
b_i	$\dfrac{25}{61}$	$\dfrac{36}{61}$	0	0
$\hat{b}_i$	$\dfrac{26}{61}$	$\dfrac{324}{671}$	$\dfrac{1}{11}$	0

order $p = 3$ (4.6.17)

SDIRK

	$\dfrac{5}{6}$				$\dfrac{5}{6}$
	$-\dfrac{15}{26}$	$\dfrac{5}{6}$			$\dfrac{10}{39}$
	$\dfrac{215}{54}$	$-\dfrac{130}{27}$	$\dfrac{5}{6}$		0
	$\dfrac{4007}{6075}$	$-\dfrac{31031}{24300}$	$\dfrac{133}{2700}$	$\dfrac{5}{6}$	$\dfrac{1}{6}$
b_i	$\dfrac{32}{75}$	$\dfrac{169}{300}$	$\dfrac{1}{100}$	0	0
$\hat{b}_i$	$\dfrac{61}{150}$	$\dfrac{2197}{2100}$	$\dfrac{19}{100}$	$-\dfrac{9}{14}$	0

order $p = 3$ (4.6.18)

For instance, a semi-implicit scheme corresponding to the optimal two-stage process (4.2.10) is given as

Table 4.5. Index $a(-b) = a \times 10^{-b}$
Numerical Results on Problem $y\prime = -y$, $y(0) = 1$
Both schemes have error order $p = 4$

h	$y(1, h)$	actual error	actual error/h^4
		Two-stage Implicit R-K Scheme (4.6.8)	
$\dfrac{1}{8}$	0.367879565	0.124054727(-6)	0.508128163(-3)
$\dfrac{1}{16}$	0.367879449	0.778600503(-8)	0.510263626(-3)
$\dfrac{1}{32}$	0.367879442	0.478114495(-9)	0.510776568(-3)
$\dfrac{1}{64}$	0.367879441	0.304521894(-10)	0.510902959(-3)
		Three-stage Semi-Implicit R-K Scheme (4.6.16)	
$\dfrac{1}{8}$	0.367879244	-0.196773630(-6)	-0.805984790(-3)
$\dfrac{1}{16}$	0.367879429	-0.119979475(-7)	-0.786297484(-3)
$\dfrac{1}{32}$	0.367879440	-0.740410705(-9)	-0.776376895(-3)
$\dfrac{1}{64}$	0.367879441	-0.459790470(-10)	-0.771400402(-3)

$$y_{n+1} = y_n + \frac{h}{4}(Y_1 + 3Y_2) ,$$

$$Y_1 = f(x_{n+1}, y_{n+1}) , \qquad\qquad (4.6.19)$$

$$Y_2 = f(x_{n+1} - \frac{2}{3}h, y_{n+1} - \frac{2}{3}hY_1) .$$

Equation (4.6.19) can be derived from the explicit scheme (4.2.10) by replacing h by $-h$.

The stability equation for the explicit two-stage R-K scheme (4.2.10) is

$$\mu_E(z) = 1+z+\frac{1}{2}z^2 \;,\; z = \lambda h \;, \tag{4.6.20}$$

which is the (2, 0) Padé approximation to e^z, while that of the semi-implicit scheme (4.6.19) is

$$\mu_S(z) = \frac{1}{1-z+\frac{1}{2}z^2} \;, \tag{4.6.21}$$

which is the (0, 2) Padé approximation to e^z.

Although (4.6.19) has a considerably large region of absolute stability, it is not L-stable, due to a rather small region of instability close to the imaginary axis.

The s-stage semi-implicit R-K scheme (of Cash 1976) can, in general, be expressed as

$$y_{n+1} = y_n + h \sum_{r=1}^{s} b_r Y_r \;, \tag{4.6.22a}$$

$$Y_r = f\,(x_n+c_r h,\; y_n+h \sum_{j=1}^{r-1} a_{rj}Y_j) \;,\; 1 \le r \le \tau \;, \tag{4.6.22b}$$

$$Y_r = f\,(x_{n+1}+c_r h,\; y_{n+1}+h \sum_{j=\tau}^{r-1} a_{rj}Y_j),\; \tau+1 \le r \le s \;. \tag{4.6.22c}$$

Equations (4.6.22a-c) contain only a set of s systems of m nonlinear equations in Y_r and is A- or L-stable provided

(a)   none of the slopes Y_r in (4.6.22b) contains a Y_r in $\tau+1 \le r \le s$, and

(b) $\tau \le [1/2\, s]$.

Cash (1976) formulated the following L-stable methods:

$$s = 4 \;,\; \tau = 1$$

$$y_{n+1} = y_n + h \sum_{r=1}^{4} b_r Y_r \,,$$

$$Y_1 = f(x_{n+1}, y_{n+1}) \,,$$

$$Y_2 = f(x_{n+1} + c_2 h, \, y_{n+1} + c_2 h Y_1) \,, \tag{4.6.23}$$

$$Y_3 = f(x_{n+1} + c_3 h, \, y_{n+1} + a_{31} h Y_1 + (c_3 - a_{31}) h Y_2) \,,$$

$$Y_4 = f(x_n, y_n) \,.$$

On obtaining the Taylor series expansion for Y_r, $r = 2, 3, 4$ about (x_{n+1}, y_{n+1}) and equating terms in powers of h with the corresponding Taylor expansion methods (4.1.1) and (4.2.2), the following relations are obtained:

$$b_1 + b_2 + b_3 + b_4 = 1 \,,$$

$$b_2 c_2 + b_3 c_3 - b_4 = -\frac{1}{2} \,,$$

$$\tag{4.6.24}$$

$$b_2 c_2^2 + b_3 c_3^2 + b_4 = \frac{1}{3} \,,$$

$$b_3 (c_3 - a_{31}) c_2 + \frac{1}{2} b_3 = \frac{1}{6} \,.$$

Two L-stable algorithms derived from (4.6.23) and (4.6.24) are

$$y_{n+1} = y_n + \frac{1}{4} h (Y_2 + 2Y_3 + Y_4) \,,$$

$$Y_4 = f(x_{n+1}, y_{n+1}) \,,$$

$$Y_2 = f(x_{n+1} - \frac{1}{3} h, \, y_{n+1} - \frac{1}{3} h Y_1) \,, \tag{4.6.25}$$

$$Y_3 = f(x_{n+1} - \frac{1}{3} h, \, y_{n+1} - \frac{1}{12} h Y_1 - \frac{1}{4} h Y_2) \,,$$

$$Y_4 = f(x_n, y_n) ,$$

whose stability function can be obtained as

$$\mu_S(z) = \frac{1 + \frac{1}{4}z}{1 - \frac{3}{4}z + \frac{1}{4}z^2 - \frac{1}{24}z^3} \qquad (4.6.26)$$

and

$$y_{n+1} = y_n + \frac{1}{6}h(Y_1 + 2Y_2 + 2Y_3 + Y_4) ,$$

$$Y_1 = f(x_{n+1}, y_{n+1}) ,$$

$$Y_2 = f(x_{n+1} - \frac{1}{2}h, \ y_{n+1} - \frac{1}{2}hY_1) , \qquad (4.6.27)$$

$$Y_3 = f(x_{n+1} - \frac{1}{2}h, \ y_{n+1} - \frac{1}{2}hY_2) ,$$

$$Y_4 = f(x_n, y_n) ,$$

with the stability function

$$\mu_S(z) = \frac{1 + \frac{1}{6}z}{1 - \frac{5}{6}z + \frac{1}{3}z^2 - \frac{1}{12}z^3} . \qquad (4.6.28)$$

The nonlinear equation

$$F(y_{n+1}) = y_{n+1} - y_n - hb_s f(x_{n+1}, y_{n+1}) - h \sum_{j=1}^{s-1} b_j Y_j = 0 , \qquad (4.6.29)$$

normally associated with this class of semi-implicit R-K processes, can be handled by the Newton methods discussed in Section 8.3. The Jacobian of the algebraic system is, unfortunately, a polynomial in the Jacobian of the ODE.

These methods have been extended by Cash and Singhal (1982), are known as MIRK methods, and are discussed in Section 9.6 .

The main attraction of implicit R-K methods is their strong stability properties. However, all such methods require the solution of systems of nonlinear equations that define the internal slopes Y_i, $i = 1(1)s$. If the R-K process is semi-implicit, then the coupled, simultaneous nonlinear system splits into s uncoupled nonlinear systems. Rosenbrock methods further reduce the problem to s uncoupled linear systems.

Further discussion on the efficient and practical implementation of implicit, semi-implicit, singly implicit, and Rosenbrock methods is found in Sections 4.7, 8.6, and 9.2.

4.7. Rosenbrock Methods

Rosenbrock (1963), in an apparent attempt to avoid the need to solve large sets of nonlinear equations with IRK and semi-implicit R-K methods, proposed the class of s-stage numerical integrators for autonomous differential systems

$$y_{n+1} = y_n + h \sum_{r=1}^{s} b_r Y_r , \tag{4.7.1a}$$

$$Y_1 = f(y_n) + d_1 h f_y(y_n) Y_1 , \tag{4.7.1b}$$

$$Y_r = f(y_n + h \sum_{j=1}^{r-1} a_{rj} Y_j)$$

$$\tag{4.7.1c}$$

$$+ d_r h f_y \left(y_n + h \sum_{j=1}^{r-1} e_{rj} Y_j \right) Y_r, \ r = 2(1)s .$$

The main advantage of the Rosenbrock method (4.7.1) is the considerable reduction in the linear algebra normally associated with the implicit and semi-implicit R-K methods, at the cost of at least one evaluation of the Jacobian at every integration step.

Wolfbrandt (1977) developed some modified Rosenbrock formulas whereby (4.7.1b,c) are replaced by

$$[I - \gamma h f_y(y_n)] Y_r = h f(y_n + \sum_{j=1}^{r-1} a_{rj} Y_j)$$

$$\tag{4.7.2}$$

$$+ hf_y(y_n) \sum_{j=1}^{r-1} e_{rj} Y_j, \; r = 1(1)s,$$

whose stability function (for order $p \geq s$) is

$$R(z) = \frac{\left[\sum_{i=0}^{s} z^i \sum_{j=0}^{i} \binom{s}{j} \frac{(-\gamma)^j}{(i-j)!} \right]}{(1-\gamma z)^s}. \tag{4.7.3}$$

Norsett and Wolfbrandt (1979) used the concept of trees to derive order relations for Rosenbrock methods.

Wanner (1980) examined the choice of optimal values of γ with regard to stability and accuracy. Kaps and Wanner (1981) suggested a more efficient form of (4.7.2) called Rosenbrock Wanner (i.e., ROW) methods and given as

$$[I - \gamma \, hf_y(y_n)] \, Y_r = hf(y_n + \sum_{j=1}^{r-1} a_{rj} \, Y_j)$$

$$\tag{4.7.4}$$

$$+ h \sum_{j=1}^{r-1} e_{rj} \, Y_j, \; r = 1(1)s.$$

Equations (4.7.2) and (4.7.4) both reduce to the original Rosenbrock scheme (4.7.1) if $e_{rj} = 0$.

Bui (1979) proposed an L-stable Rosenbrock method of order four with a built-in error estimator. Kaps and Rentrop (1979) also obtained a ROW method of order four, using an embedded third-order formula as the error estimator. Kaps and Wanner (1981) derived higher order ROW methods (i.e., $p=5,6$), using just one Jacobian evaluation per integration step.

Steihaug and Wolfbrandt (1979) extended the Rosenbrock formulas (4.7.1) to the form

$$[I - h \, e_{rr} \, A] \, Y_r = f(y_n + h \sum_{j=1}^{r-1} a_{rj} \, Y_j)$$

$$\tag{4.7.5}$$

$$+ h \, A \sum_{j=1}^{r-1} e_{rj} \, Y_j, \; r = 1(1)s,$$

where A is no longer the exact Jacobian, but a real square matrix of order m, and h is chosen to ensure that the iteration matrix $I - h \, e_{rr} A$ is nonsingular.

The generalized Rosenbrock method for the IVP (2.1.4) is described as

$$(I - \gamma \, hf_y(x_n, y_n)) \, Y_r = f(x_n + c_r h, \; y_n + \sum_{j=1}^{r-1} a_{rj} \, Y_j) + d_r hf_x(x_n, y_n)$$

$$+ \sum_{j=1}^{r-1} e_{rj} Y_j , \ r = 1(1)s, \tag{4.7.6a}$$

$$y_{n+1} = y_n + h \sum_{r=1}^{s} b_r Y_r, \tag{4.7.6b}$$

where

$$d_1 = \gamma, \ d_r = \gamma + \sum_{j=1}^{r-1} e_{rj} d_j , \ r = 2(1)s , \tag{4.7.6c}$$

$$c_r = \sum_{j=1}^{r-1} a_{rj} d_j / \gamma , \ r = 1(1)s . \tag{4.7.6d}$$

Shampine (1982) wrote a mathematical software based on the ROW method GRK4A of Kaps and Rentrop (1979), given by (4.7.7) for a stiff step, and on the Fehlberg (1969) (4,5) pair, given by (4.4.14).

The GRK4A pair for the non-autonomous IVPs (2.1.4) is described as

$$W_n = I - \frac{1}{2} h f_y(x_n, y_n) , \tag{4.7.7a}$$

$$W_n Y_1 = f (x_n, y_n) + \frac{1}{2} h f_x(x_n, y_n) , \tag{4.7.7b}$$

$$W_n Y_2 = f (x_n + h, y_n + h Y_1) - \frac{3}{2} h f_x(x_n, y_n) - 4 Y_1 , \tag{4.7.7c}$$

$$W_n Y_3 = f (x_n + \frac{3}{5} h, y_n + \frac{24}{25} h Y_1 + \frac{3}{25} h Y_2) + \frac{121}{50} h f_x(x_n, y_n)$$
$$+ \frac{186}{25} Y_1 + \frac{6}{5} Y_2 , \tag{4.7.7d}$$

$$W_n Y_4 = f (x_n + \frac{3}{5} h, y_n + \frac{24}{25} h Y_1 + \frac{3}{25} h Y_2) + \frac{29}{250} h f_x(x_n, y_n)$$
$$- \frac{56}{125} Y_1 - \frac{27}{125} Y_2 - \frac{1}{5} Y_3 , \tag{4.7.7e}$$

$$y_{n+1,3} = y_n + h(\frac{98}{108} Y_1 + \frac{11}{72} Y_2 + \frac{25}{16} Y_3) , \tag{4.7.7f}$$

$$y_{n+1,4} = y_n + h(\frac{19}{18} Y_1 + \frac{1}{4} Y_2 + \frac{25}{216} Y_3 + \frac{125}{216} Y_4) , \tag{4.7.7g}$$

$$est = y_{n+1,4} - y_{n+1,3} = h(\frac{17}{108} Y_1 + \frac{7}{72} Y_2 + \frac{125}{216} Y_4) .$$ (4.7.7h)

Shampine (1982) derived a cheap and effective estimate of the conditioning of the linear system. The scaled iteration matrix

$$\overline{W}_n = f_y - \frac{1}{h \gamma} I$$ (4.7.8)

was adopted.

The Fehlberg (4,5) option in DEGRK effects local extrapolation, and, thus, is effectively of order five. The ROW option only does four solutions of linear systems at every integration step, and the stepsize is restricted to ensure that

$$h \gamma f_y \leq 10^{10} .$$ (4.7.9)

5. LINEAR MULTISTEP METHODS

5.1. Starting Procedure

Consider the initial value problem (2.1.4), which satisfies the hypothesis of the Existence and Uniqueness Theorem (Theorem 2.1). The sequence $\{x_n \mid x_n = a + nh, \, n > 0\}$ constitutes a discrete point set defined on the integration interval $[a, b]$ where $h \geq 0$ is a suitable meshsize, possibly allowed to vary in accordance with certain desired criteria. Except in Section 5.6, h is assumed to be fixed.

Provided the points $\{(x_{n+j}, y_{n+j}) \mid j = 0(1)k-1\}$ belong to the solution space of the IVP (2.1.4), the k-step linear multistep method (LMM) "extrapolates" the next point in the solution space with the following k-th order finite difference equation:

$$\rho(E)y_n = h\sigma(E)f_n, \, f_n = f(x_n, y_n), \tag{5.1.1a}$$

where E is the shift operator defined as $E^j y_n = y_{n+j}$, and $\rho(E)$ and $\sigma(E)$ are, respectively, the first and second characteristic polynomials of the LMM. These polynomials are defined respectively as

$$\rho(E) = \sum_{j=0}^{k} \alpha_j E^j, \, \alpha_k \neq 0 \tag{5.1.1b}$$

and

$$\sigma(E) = \sum_{j=0}^{k} \beta_j E^j, \tag{5.1.1c}$$

usually with the added assumption that $\alpha_0^2 + \beta_0^2 > 0$.

Degree (σ) < degree (ρ) implies that the method is explicit, and the unknown quantity y_{n+k} in (5.1.1) can be obtained directly from the solution of the difference equation. Otherwise, if (5.1.1) is implicit, iterative processes have to be adopted. This introduces the problems of generating a good initial estimate and of ascertaining conditions which ensure the convergence of the iterative process.

A desired solution is generated from (5.1.1), provided the past values $\{y_n, y_{n+1}, ..., y_{n+k-1}\}$ are available. Unfortunately, at the initial stage, only the starting value y_0 is known. Hence, we cannot apply the LMM (5.1.1) except when the necessary additional starting values $y_1, y_2, ..., y_{k-1}$ are available. These values can be generated with a Runge-Kutta method whose order of accuracy is at least as high as that of the LMM under consideration. Explicit R-K schemes are recommended for the nonstiff phase, while the implicit R-K schemes might be adopted in the stiff phase. The variable order variable step methods are self-starting. Explicit methods are often feasible for variable stepsize. With Adams methods and Backward Differentiation Formulas, orders two and one are respectively adopted to generate the starting values.

The automatic numerical integrators (LMM) could generate each of the necessary additional starting values y_j (for $1 \leq j \leq k-1$) by a j-step LMM that is one of the methods. The variable order mechanism is incorporated, since, in general, the order of the LMM increases with the stepnumber.

Recently, Gear (1980d) used an alternative approach for stiff IVPs earlier mentioned by Ceschino and Kuntzmann (1966).

The $(k-1)$ additional starting values are generated by the single step $(k-1)$-stage R-K process of the form

$$y_j = y_0 + h \sum_{r=1}^{k-1} \gamma_{jr} Y_r \,, \; j = 1(1)k-1 \,, \tag{5.1.2a}$$

with the slopes Y_r given as

$$Y_r = hf\left(x_0 + h\alpha_r, \; y_0 + h\sum_{s=1}^{k-1} \beta_{rs} Y_s\right), \; r = 1(1)k-1 \,, \tag{5.1.2b}$$

with the constraint

$$\alpha_r = \sum_{s=1}^{k-1} \beta_{rs} \,. \tag{5.1.2c}$$

Gear proposed a fourth order six-stage explicit R-K process for nonstiff IVPs, while he proposed $(k-1)$-th order $(k-1)$-stage R-K process to generate the $k-1$ points simultaneously.

In case the LMM (5.1.1) is implicit, it is necessary to solve the nonlinear algebraic equation

$$F(y_{n+k}) = y_{n+k} - h\beta_k f(x_{n+k}, y_{n+k}) + \sum_{j-0}^{k-1} [\alpha_j y_{n+j} - h\beta_j f_{n+j}] = 0, \qquad (5.1.3)$$

which may be solved by one point iteration scheme for nonstiff methods, and by a modified Newton sceme for a stiff method.

Most LMMs (5.1.1) are formulated by one of the following approaches:

(a) method of undetermined coefficients,

(b) numerical differentiation, or

(c) numerical integration.

Methods based on the numerical integration approach provide avenues for error estimation and error analysis, which, in effect, facilitate stepsize adjustment.

5.2. Explicit Linear Multistep Methods (Adams-Bashforth 1883, and related methods)

The linear multistep method is essentially a polynomial interpolation procedure whereby the solution or its derivative is replaced by a polynomial of appropriate degree in the independent variable x, whose derivative or integral is readily obtained. This accounts for the poor performance of the LMM when dealing with IVPs (2.1.4) whose solutions contain singularities, or those with large Lipschitz constants, as polynomial approximations are inappropriate for these classes of problems.

The theoretical solution $y(x)$ to the IVP (2.1.4) satisfies the following integral equation

$$y(x_{n+1}) = y(x_n) + \int_{x_n}^{x_{n+1}} f(\tau, y(\tau))d\tau, \qquad (5.2.1)$$

whose integrand $f(x, y(x))$ can be adequately represented by a unique polynomial $p_{k-1}(x)$ of maximum degree $k-1$, which passes through the set of points $\{(x_{n-j}, f_{n-j}) \mid j = 0(1)k-1\}$ and is given by

$$p_{k-1}(x) = f_n + (x-x_n)\frac{\nabla f_n}{h} + (x-x_n)(x-x_{n-1})\frac{\nabla^2 f_n}{2!h^2}$$

$$\qquad (5.2.2)$$

$$+ ... + (x-x_n)(x-x_{n-1})...(x-x_{n-k+2})\frac{\nabla^{k-1} f_n}{(k-1)!h^{k-1}}.$$

The error

$$E(x) = p_{k-1}(x) - y'(x)$$

of the interpolating polynomial was established in Theorem 1.2 as

$$E(x) = \frac{\prod_{i=0}^{k-1}(x-x_{n-i})y^{(k+1)}(\xi)}{k!} \tag{5.2.3}$$

for some $\xi \, \varepsilon [x_{n-k+1}, x_{n-k,...}, x_n]$. Further detail on interpolation theory can be found in Atkinson (1978) and Shampine and Allen (1973).

By inserting (5.2.2) in (5.2.1), a numerical approximation y_{n+1} to $y(x_{n+1})$ is given as

$$y_{n+1} = y_n + \int_{x_n}^{x_{n+1}} [f_n + (x-x_n)\frac{\nabla f_n}{h} + (x-x_n)(x-x_{n-1})\frac{\nabla^2 f_n}{2!h^2}$$

$$\tag{5.2.4}$$

$$+ ... + (x-x_n)(x-x_{n-1})...(x-x_{n-k+2})\frac{\nabla^{k-1}f_n}{(k-1)!h^{k-1}}]dx \, .$$

The transformation

$$s = \frac{x-x_n}{h} \tag{5.2.5}$$

reduces (5.2.4) to the form

$$y_{n+1} - y_n = h\sum_{r=0}^{k-1}\gamma_r\nabla^r f_n \, , \tag{5.2.6a}$$

whose coefficients γ_r are specified by

$$\gamma_r = (-1)^r\int_0^1 \binom{-s}{r}ds \, . \tag{5.2.6b}$$

Exercise. Show that $|\gamma_r| < 1$ for all $r > 0$.

The generating function $G(t)$, defined by

$$G(t) = \sum_{r=0}^{\infty}\gamma_r t^r \, , \tag{5.2.7}$$

can be used to generate the coefficients γ_r recursively, in the spirit of Henrici (1962). The summation in (5.2.7) is absolutely convergent, provided $-1 < t < 1$, since, from (5.2.6b), each γ_r is bounded above by unity.

From (5.2.6b) and (5.2.7), we obtain the generating function $G(t)$ as

$$G(t) = \sum_{r=0}^{\infty}[(-1)^r\int_0^1\binom{-s}{r}ds]t^r$$

$$= \sum_{r=0}^{\infty} (-t)^r \int_0^1 \binom{-s}{r} ds$$

$$= \int_0^1 \sum_{r=0}^{\infty} (-t)^r \binom{-s}{r} ds \; ;$$

i.e.,

$$G(t) = \int_0^1 (1-t)^{-s} ds \; . \tag{5.2.8}$$

The insertion of the identity

$$1 - t = e^{\log_e(1-t)}$$

into (5.2.8) implies

$$G(t) = \int_0^1 e^{-s\log_e(1-t)} ds \; ,$$

which is integrated to get

$$G(t) = -\frac{1}{\log_e(1-t)} \, [e^{-s\log_e(1-t)}]_0^1$$

$$= -\frac{1}{\log_e(1-t)} \, [\frac{1}{1-t} - 1] \; .$$

This implies

$$G(t) = -\frac{t}{(1-t)\log_e(1-t)} \; . \tag{5.2.9}$$

Equation (5.2.9) can be rearranged as

$$-\frac{\log_e(1-t)}{t} G(t) = \frac{1}{1-t} \; . \tag{5.2.10}$$

Adopting the expansions

$$-\frac{\log_e(1-t)}{t} = \sum_{r=0}^{\infty} \frac{t^r}{r+1} \tag{5.2.11}$$

and

$$\frac{1}{1-t} = \sum_{r=0}^{\infty} t^r \tag{5.2.12}$$

in (5.2.10) leads to the following identity:

$$(\sum_{r=0}^{\infty} \frac{t^r}{r+1})(\sum_{r=0}^{\infty} \gamma_r t^r) = \sum_{r=0}^{\infty} t^r . \tag{5.2.13}$$

On equating the coefficients of powers of t in (5.2.13), we can recursively generate the coefficients of the integration formula (5.2.6) as

$$\sum_{r=0}^{k} \frac{1}{r+1} \gamma_{k-r} = 1 , k = 0, 1, 2,.... \tag{5.2.14}$$

The integration formula (5.2.6) with coefficients γ_r given by (5.2.14) is the well-known Adams-Bashforth method proposed by Bashforth (1883), whose coefficients are given in Table 5.1.

The local truncation error t_{n+1} in the Adams-Bashforth method (5.2.6), which is due to the interpolation error $E(x)$ given by (5.2.3), can be expressed as

$$t_{n+1} = \int_{x_n}^{x_{n+1}} E(x)dx$$

$$\tag{5.2.15}$$

$$= \int_{x_n}^{x_{n+1}} \prod_{i=0}^{k-1} \frac{(x-x_{n-i})}{k!} f^{(k)}(\xi)dx .$$

The adoption of the transformation (5.2.5) in (5.2.15) yields the following estimate for the local truncation error for (5.2.6):

$$t_{n+1} = h^{k+1} \int_{0}^{1} \prod_{i=0}^{k-1} \frac{(s+i)}{k!} f^{(k)}(\xi)ds$$

Table 5.1. Coefficients of Adams-Bashforth Formula (5.2.6)

k	0	1	2	3	4
γ_k	1	$\dfrac{1}{2}$	$\dfrac{5}{12}$	$\dfrac{3}{8}$	$\dfrac{251}{720}$

5	6	7	8	9	10
$\dfrac{95}{288}$	$\dfrac{19087}{60480}$	$\dfrac{5257}{17280}$	$\dfrac{1070017}{3628800}$	$\dfrac{25713}{89600}$	$\dfrac{26842253}{95800320}$

$$= h^{k+1} \int_0^1 (-1)^k (\frac{-s}{k}) f^{(k)}(\xi) ds \ . \tag{5.2.16}$$

The influence function (cf. 5.5.16) approach can be applied to bound the righthand side of (5.2.16).

An alternative expression for the integration formula (5.2.6a) can be obtained by inserting the identity

$$\nabla^r f_n = \sum_{j=0}^{r} (-1)^j \binom{r}{j} f_{n-j} \tag{5.2.17}$$

in (5.2.6a). This yields

$$y_{n+1} - y_n = h \sum_{j=0}^{k} \gamma_{kj} f_{n-j} \ , \tag{5.2.18}$$

with

$$\gamma_{kj} = (-1)^{j-1} \sum_{i=j-1}^{k-1} \binom{i}{j-1} \gamma_i \ . \tag{5.2.19}$$

The Adams-Bashforth formulas are really captivating because, being explicit, they involve a straightforward computation of y_{n+1}. However, it will be shown later that such procedures are handicapped by lower attainable order compared with the corresponding implicit formulas. Besides, their stability is quite inferior to that of the corresponding implicit process. In practical applications, the explicit Adams are used to generate the necessary additional starting values $y_1, y_2, ..., y_{k-1}$, respectively, at $x_1, x_2, ..., x_{k-1}$ for their corresponding implicit formulas, whose development will be treated in the next section.

5.3. Implicit Linear Multistep Methods (Adams-Moulton scheme, 1926 and related methods)

The procedure for the development of the implicit linear multistep methods is identical to that of the explicit linear multistep methods—the basic difference being the range of integration.

For the implicit schemes corresponding to (5.2.6), Equation (5.2.1) is replaced by the integral equation

$$y(x_n) = y(x_{n-1}) + \int_{x_{n-1}}^{x_n} f(\tau, y(\tau)) d\tau \ . \tag{5.3.1}$$

Following the same arguments as for the Adams-Bashforth formulas, the Adams-Moulton formulas proposed by Moulton (1926) can be derived from (5.3.1) as

Table 5.2. Coefficients of Adams-Moulton Methods (5.3.2)

k	0	1	2	3	4
γ_k^*	1	$-\dfrac{1}{2}$	$-\dfrac{1}{12}$	$-\dfrac{1}{24}$	$-\dfrac{19}{720}$

5	6	7	8	9	10
$-\dfrac{3}{160}$	$-\dfrac{863}{60480}$	$-\dfrac{275}{24792}$	$-\dfrac{33953}{3628800}$	$-\dfrac{8183}{1036800}$	$-\dfrac{3250433}{479001600}$

$$y_n - y_{n-1} = h \sum_{r=0}^{k} \gamma_r^* \nabla^r f_n \, , \tag{5.3.2}$$

with the coefficients γ_r^* given by

$$\gamma_r^* = (-1)^r \frac{1}{h} \int_{x_{n-1}}^{x_n} \binom{-s}{r} ds$$

$$\tag{5.3.3}$$

$$= (-1)^r \int_{-1}^{0} \binom{-s}{r} ds \, .$$

The coefficients γ_r^* can be generated recursively with the generating function

$$G^*(t) = \sum_{r=0}^{\infty} \gamma_r^* t^r \, , \tag{5.3.4}$$

whose coefficients γ_r^* can be obtained recursively as

$$\gamma_0^* = 1 \, ,$$

$$\tag{5.3.5}$$

$$\sum_{r=0}^{k} \frac{1}{r+1} \gamma_{k-r}^* = 0 \, , \, k > 1 \, .$$

A comparison of Table 5.1 and Table 5.2 reveals that the coefficients of the implicit formula are smaller than those of the corresponding explicit formulas. The smaller coefficients lead to smaller local truncation errors and, hence, to improved accuracy over the explicit Adams-Bashforth methods.

As a matter of fact, the generating functions $G(t)$ for the Adams-Bashforth method, given by (5.2.7), and $G^*(t)$ for the Adams-Moulton scheme, given by (5.3.4), are related as follows:

$$G(t) = (1-t)^{-1} G^*(t) .$$ (5.3.6)

Exercise.

1. Deduce from (5.3.6) that

$$\gamma_k = \sum_{r=0}^{k} \gamma_r^* , \quad k \geq 0 .$$ (5.3.7)

2. Derive a k-step explicit formula from the integral equation

$$y(x_{n+1}) - y(x_{n-2}) = \int_{x_{n-2}}^{x_{n+1}} f(\tau, y(\tau)) d\tau ,$$ (5.3.8)

   and the equivalent corrector formula from

$$y(x_n) - y(x_{n-3}) = \int_{x_{n-3}}^{x_n} f(\tau, y(\tau)) d\tau .$$ (5.3.9)

Establish the identities (5.3.6) and (5.3.7), using the generating functions of the resultant LMM from (5.3.8) and (5.3.9).

Henrici (1962) also discusses the explicit scheme of Nystrom (1925), whose development is based on the integral equation

$$y(x_{n+1}) - y(x_{n-1}) = \int_{x_{n-1}}^{x_{n+1}} f(\tau, y(\tau)) d\tau .$$ (5.3.10)

This results in the explicit integration formula

$$y_{n+1} - y_{n-1} = h \sum_{r=0}^{k-1} \lambda_r \nabla^r f_n ,$$ (5.3.11)

whose coefficients λ_r can be obtained from

$$\sum_{r=0}^{k} \frac{1}{r+1} \lambda_{k-r} = \begin{cases} 2 , & k = 0 , \\ 1 , & k \geq 1 . \end{cases}$$ (5.3.12)

The Milne procedures are the corresponding implicit LMMs for the Nystrom formulas. They are derived from the integral equation

$$y(x_n) - y(x_{n-2}) = \int_{x_{n-2}}^{x_n} f(\tau, y(\tau)) d\tau .$$ (5.3.13)

This yields the difference equation

$$y_n - y_{n-2} = h \sum_{r=0}^{k} \lambda_r^* \nabla^r f_n ,$$ (5.3.14)

whose coefficients can be obtained as

Table 5.3. Coefficients of Nystrom Formulas (5.3.11)

k	0	1	2	3	4
λ_k	2	0	$\dfrac{1}{3}$	$\dfrac{1}{3}$	$\dfrac{29}{90}$

5	6	7	8	9	10
$\dfrac{14}{45}$	$\dfrac{1139}{3980}$	$\dfrac{41}{140}$	$\dfrac{32377}{113400}$	$\dfrac{3956}{14175}$	$\dfrac{3046263}{7484400}$

$$\lambda_0^* = 2 \ ,$$

$$\lambda_1^* + \frac{1}{2}\lambda_0^* = -2 \ , \tag{5.3.15}$$

$$\sum_{r=0}^{k}\frac{1}{r+1}\lambda_r^* = 0 \ , \ k \geq 2 \ .$$

As a matter of fact, the Relations (5.3.6) and (5.3.7) also hold for the coefficients of the Nystrom-Milne predictor-corrector pairs (5.3.11) and (5.3.14).

Apart from the better stability of the predictor-corrector formulas (introduced in Section 5.4) over the explicit formulas, the predictor-corrector formulas are generally more accurate and provide reasonable and adequate error estimators.

Numerical Differentiation Approach

Table 5.4. Coefficients of Milne's Methods

k	0	1	2	3	4
λ_k^*	2	-2	$\dfrac{1}{3}$	0	$-\dfrac{1}{90}$

5	6	7	8	9	10
$-\dfrac{1}{90}$	$-\dfrac{37}{3780}$	$-\dfrac{8}{945}$	$-\dfrac{119}{16200}$	$\dfrac{9}{1400}$	$\dfrac{8501}{1496880}$

In the previous two sections, we have discussed LMMs derived by the replacement of the derivatives $f(x, y)$ (r.h.s. of (2.1.4)) by a polynomial of suitable degree and subsequent integration. An alternative strategy would be to replace the theoretical solution $y(x)$ (l.h.s. of (2.1.4)) by an appropriate polynomial $P_k(x)$ and then differentiate. Although this approach does not provide a basis for good error estimation, the resultant numerical integrators (backward differentiation formulas) are well-suited to stiff initial value problems.

On rewriting the IVP (2.1.4) as

$$hDy(x_{n+k}) = hf(x_{n+k}, y(x_{n+k})) ,$$ (5.3.16)

with $D = d/dx$, and using the identity

$$hD = -\log_e(1-\nabla) = \sum_{r=0}^{k} \frac{\nabla^{r+1}}{r+1} ,$$ (5.3.17)

we obtain the following truncated difference equation of order k:

$$\sum_{r=0}^{k} \lambda_r y_{n-r+k} = hf_{n+k} .$$ (5.3.18)

For example, for $k = 0$, (5.3.16) gives

$$\nabla y_n = hf_n ;$$

i.e.,

$$y_n - y_{n-1} = hf_n , \text{ or}$$

(5.3.19)

$$y_{n+1} - y_n = hf_{n+1} , p = 1 .$$

which is the implicit Euler scheme (3.2.7).

With $k = 1$, (5.3.16) gives

$$(\nabla + \frac{1}{2}\nabla^2)y_{n+1} = hf_{n+1} ;$$

i.e.,

$$y_{n+2} - \frac{4}{3}y_n + \frac{1}{3}y_{n-1} = \frac{2}{3}hf_{n+1}$$

or

$$y_{n+2} - \frac{4}{3}y_{n+1} + \frac{1}{3}y_n = \frac{2}{3}hf_{n+2} , p = 2 .$$ (5.3.20)

The coefficients λ_r for the BDF of step number $k \le 6$ are given in Table 5.5.

Another alternative to the derivation of LMM is the method of undetermined coefficients, which was illustrated earlier in the derivation of the trapezoidal scheme in Section 3.4.

Cryer (1972, 1973) and Hairer and Wanner (1984) established that, for constant meshsize, the BDF are zero-stable if and only if the step number $k \leq 6$.

5.4. Implementation of the Predictor-Corrector Formulas

In the implementation of the implicit linear multistep methods, there are three distinct stages:

Stage 1. **Predict** the starting value for the dependent variable y_{n+k} as $y_{n+k}^{[0]}$; i.e.,

$$\text{P:} \quad y_{n+k}^{[0]} = \sum_{j=0}^{k-1} (-\alpha_{1j} y_{n+j} + \beta_{1j} h y_{n+j}') \, . \tag{5.4.1}$$

Stage 2. This is simply the **Evaluation** of the derivative at $(x_{n+k}, y_{n+k}^{[0]})$; i.e.

$$\text{E:} \quad f_{n+k}^{[0]} = f(x_{n+k}, y_{n+k}^{[0]}) \, . \tag{5.4.2}$$

Stage 3. This is a **Correction** represented as

$$\text{C:} \quad y_{n+k}^{[1]} = \sum_{j=0}^{k-1} (-\alpha_{2j} y_{n+j} + \beta_{2j} h y_{n+j}') + \beta_{2k} h f_{n+k}^{[0]} \, . \tag{5.4.3}$$

A combination of the three stages is called PEC mode. It is often more desirable in terms of stability considerations to incorporate one additional

Table 5.5. Coefficients of Backward Differentiation Formula (5.3.18)

k	λ_0	λ_1	λ_2	λ_3	λ_4	λ_5	λ_6
1	1	-1					
2	$\dfrac{3}{2}$	-2	$\dfrac{1}{2}$				
3	$\dfrac{11}{6}$	-3	$\dfrac{3}{2}$	$-\dfrac{1}{3}$			
4	$\dfrac{25}{12}$	-4	3	$-\dfrac{4}{3}$	$\dfrac{1}{4}$		
5	$\dfrac{137}{60}$	-5	5	$-\dfrac{10}{3}$	$\dfrac{5}{4}$	$-\dfrac{1}{5}$	
6	$\dfrac{147}{60}$	-6	$\dfrac{15}{2}$	$-\dfrac{20}{3}$	$\dfrac{15}{4}$	$-\dfrac{6}{5}$	$\dfrac{1}{6}$

function evaluation per integration step,

$$\text{E:} \quad f_{n+k}^{[1]} = f(x_{n+k}, y_{n+k}^{[1]}), \tag{5.4.4}$$

thus yielding the PECE mode.

There are two options available in the implementation of any predictor-corrector formula:

a. The first option is to repeat stages two and three until a certain prescribed convergence criterion is satisfied; i.e.,

$$\text{E:} \quad f_{n+k}^{[r]} = f(x_{n+k}, y_{n+k}^{[r]}),$$

$$\tag{5.4.5}$$

$$\text{C:} \quad y_{n+k}^{[r+1]} = \sum_{j=0}^{k-1} (-\alpha_{2j} y_{n+j} + \beta_{2j} h y'_{n+j}) + \beta_{2k} h y_{n+k}^{[r]}$$

for $r = 0, 1, 2,....$

The ultimate numerical solution is denoted by y_{n+k}. This mode is called "iterating to convergence." Its stability properties are essentially those of the corrector in as far as the stability region of the corrector is inside the convergence circle with radius $1/\beta_{2k}$.

b. The other option is to repeat stages two and three for a fixed number of times, μ, which yields $y_{n+k} = y_{n+k}^{[\mu]}$. In practical applications, μ is normally chosen as 1 or 2, with the tacit assumption that the predictor formula is of sufficiently high order q for $p-1 \le q \le p$, where p is the order of the corrector formula. The entire process can be implemented in either the $P(EC)^\mu$ or $P(EC)^\mu E$ mode, the latter being acclaimed to possess superior stability characteristics (cf. Hull and Creemer 1963). The two modes are explicitly described as follows:

$$\text{P:} \quad y_{n+k}^{[0]} = \sum_{j=0}^{k-1} (-\alpha_{1j} y_{n+j}^{[\mu]} + \beta_{1j} h f_{n+j}^{[v]}),$$

$$\text{E:} \quad f_{n+k}^{[r]} = f(x_{n+k}, y_{n+k}^{[r]}), \tag{5.4.6}$$

$$\text{C:} \quad y_{n+k}^{[r+1]} = \sum_{j=0}^{k-1} (-\alpha_{2j} y_{n+j}^{[\mu]} + \beta_{2j} h f_{n+j}^{[v]}) + \beta_{2k} h f_{n+k}^{[r]},$$

for $r = 0(1)\mu-1$;

$$\text{E:} \quad f_{n+k}^{[\mu]} = f(x_{n+k}, y_{n+k}^{[\mu]}),$$

if operating in the $P(EC)^\mu E$ mode; and

$$v = \begin{cases} \mu-1 & \text{for the } P(EC)^\mu\text{-mode}, \\ \mu & \text{for the } P(EC)^\mu E\text{-mode}. \end{cases} \tag{5.4.7}$$

In the event that the predictor and the corrector are of the same order p, Milne (1962) gives a heuristic reasoning of the following error estimates:

$$t_{n+1}^p \equiv C_{1p+1} h^{p+1} y^{(p+1)}(x_n)$$

(5.4.8)

$$= \frac{C_{1p+1}}{(C_{1p+1}-C_{2p+1})} (y_{n+k}^{[\mu]} - y_{n+k}^{[0]})$$

for the predictor formula, and

$$t_{n+1}^C \equiv C_{2p+1} h^{p+1} y^{(p+1)}(x_n)$$

(5.4.9)

$$= \frac{C_{2p+1}}{(C_{1p+1} - C_{2p+1})} (y_{n+k}^{[\mu]} - y_{n+k}^{[0]})$$

for the corrector formula, where C_{1p+1} and C_{2p+1} are the error constants for both the predictor and the corrector, respectively.

Example. The numerical solution to the initial value problem

$$y_1' = y_2, \, y_1(0) = 0 \, , 0 \le x \le 3\pi/2 \, ,$$

$$y_2' = -y_1 \, , \, y_2(0) = 1$$

was generated with a uniform meshsize $h = \pi/16$, using both the $P(EC)^\mu$- and $P(EC)^\mu E$- modes for $\mu = 1, 2$. The following schemes were adopted:

Method 1. Adams-Bashforth-Moulton (5.2.6) and (5.3.2) with $k = 4$ and 3, respectively.

Predictor:

$$y_{n+4} = y_{n+3} + \frac{h}{24}(-9f_n + 37f_{n+1} - 59f_{n+2} + 55f_{n+3}) \, ,$$

(5.4.10)

$$C_{15} = \frac{251}{720} \, .$$

Corrector:

$$y_{n+4} = y_{n+3} + \frac{h}{24}(f_{n+1} - 5f_{n+2} + 19f_{n+3} + 9f_{n+4}) \, ,$$

(5.4.11)

$$C_{25} = -\frac{19}{720} \, .$$

Method 2. LMM Derived from (5.3.8) and (5.3.9) with stepnumber $k = 4$ and 3, respectively.

Predictor:

$$y_{n+4} = y_{n+1} + \frac{h}{8}(-3f_n + 15f_{n+1} - 9f_{n+2} + 21f_{n+3}),$$

(5.4.12)

$$C_{15} = \frac{27}{80}.$$

Corrector:

$$y_{n+4} = y_{n+1} + \frac{h}{8}(3f_{n+1} + 9f_{n+2} + 9f_{n+3} + 3f_{n+4}),$$

(5.4.13)

$$C_{25} = -\frac{3}{80}.$$

It can be observed that the two Correctors (5.4.11) and (5.4.13) are effectively three-step schemes, as $\alpha_{20} = \beta_{20} = 0$. This is, however, tolerable, since the ultimate goal is to ensure that the error order of the predictor correspond with that of the corrector in order to be able to adopt Milne's error estimators (5.4.8) and (5.4.9).

It is observed that the error constants for the correctors are, in general, smaller than those of the predictors.

The details of the numerical experiments are given in Tables 5.6a, 5.6b, 5.7a, and 5.7b. It is observed that no significant advantage is derived by adopting more than one iteration for this particular problem and meshsize h. As expected, the global (exact) errors are close to the local error estimates near the starting values, and, as integration progresses, the global errors are larger than the local error estimates.

The Milne error estimates, though simple and efficient, are not valid in general; the appropriate generalization, independently discovered by Stetter (1973, pages 260 and 294) and by Gear (1974), was

$$d_{n+k} = \frac{K(y_{n+k} - y_{n+k}^{[0]})}{h},$$

(5.4.14)

where

$$K = \frac{\left(\dfrac{a_k^*}{\rho_1'(1)}\right) \dfrac{C_{2p+1}}{\rho_2'(1)}}{\dfrac{C_{2p+1}}{\rho_2'(1)} - \dfrac{C_{1p+1}}{\rho_1'(1)}}$$

(5.4.15)

(ρ_1 and ρ_2 being the first characteristic polynomials of both the predictor and the corrector, respectively).

Lambert (1971) established that every predictor-corrector pair P, C has a corresponding predictor $\overline{P}$ such that $P(EC)^\mu$ mode is computationally equivalent to $\overline{P}(EC)^{\mu-1}E$ mode and possesses identical stability properties.

5.5. General Theory of Linear Multistep Methods

All the linear multistep formulas discussed in Sections 5.2 and 5.3 fall under the umbrella of the general k-step linear multistep methods

$$y_r = S_r(h) , \ 0 \le r \le k-1 \text{ (starting values)} , \tag{5.5.1a}$$

$$\sum_{j=0}^{k} \alpha_j y_{n+j} = h \sum_{j=0}^{k} \beta_j f_{n+j} , \ f_{n+j} = f(x_{n+j}, y_{n+j}) , \tag{5.5.1b}$$

whose coefficients α_j, β_j; $j = 0(1)k$ are presumed to be real and satisfy the constraints $\alpha_k \ne 0$, $\alpha_0^2 + \beta_0^2 > 0$. A violation of the last constraint indicates the existence of an equivalent LMM with smaller stepnumber. There is no loss of generality in normalizing the coefficients of (5.5.1) to ensure that $\sigma(1) = \sum_{j=0}^{k}\beta_j = 1$.

The integration formula (5.5.1) is said to be linear, as the r.h.s. is a linear combination of the derivatives $f_{n+j}, j = 0(1)k$.

In the remainder of this section, we identify the minimum properties that (5.5.1) should possess in order to guarantee adequate stability and convergence characteristics, thus ensuring that acceptable results are attainable.

We associate (5.5.1) with a linear difference operator $L[y(x), h]$, given as

$$L[y(x_n), h] = \sum_{j=0}^{k}[\alpha_j y(x_n+jh) - h\beta_j f(x_n+jh, y(x_n+jh))] , \tag{5.5.2}$$

where $y(x)$ is a sufficiently differentiable function. Equation (5.5.2) defines the local truncation error t_{n+k} of the LMM (5.5.1) if $y(x)$ represents a solution to (2.1.4).

If $y(x)$ is at least $(p+1)$ times continuously differentiable, the use of appropriate Taylor expansions of $\{y(x_{n+j})$ and $y'(x_{n+j})|j = 0(1)k\}$ about a suitable meshpoint $x_{n+\nu}, 0 \le \nu \le k$ reduces (5.5.2) to the form

$$L[y(x_n), h] = c_0 y(x_{n+\nu}) + c_1 hy'(x_{n+\nu}) + ...$$

$$+ ... + c_r h^r y^{(r)}(x_{n+\nu}) + ..., \tag{5.5.3}$$

where the constants $c_r, r = 0(1)k$ are given as follows:

$$c_0 = \sum_{j=0}^{k}\alpha_j ,$$

Table 5.6a. Adams-Bashforth-Moulton (5.4.10) and (5.4.11)
Linear Multistep Method of Order Four
Number of Differential Equations = 2
Predictor Corrector Constant = 3.486111111d-01
Corrector error constant = -2.638888889d-02

P(EC)	Mode
Uniform Stepsize =	1.963495408d-01

Starting Values

x	y
0.000000000d-01	0.000000000d-01
	1.000000000d-00
1.9634595408d-01	1.950903220d-01
	9.807852804d-01
3.926990817d-01	3.826834324d-01
	9.238795325d-01
5.890486225d-01	5.555702330d-01
	8.314696123d-01

	Numerical Solution		Milne's Error Estimates		
x	y		Predictor	Corrector	Exact Error
7.85398d-01	7.07116d-01		9.59750d-05	-7.26504d-06	9.38799d-06
	7.07110d-01		-3.22605d-05	2.44203d-06	3.01723d-06
9.81748d-01	8.31498d-01		8.06067d-05	-6.10170d-06	2.83791d-05
	5.55587d-01		-7.78654d-05	5.89419d-06	1.67927d-05
1.17810d-00	9.23932d-01		6.41292d-05	-4.85440d-06	5.21278d-05
	3.82701d-01		-4.77518d-05	3.61468d-06	1.71350d-05
1.37445d-00	9.80856d-01		7.58001d-05	-5.73786d-06	7.02922d-05
	1.95102d-01		-8.96377d-05	6.78533d-06	1.20552d-05
1.57080d-00	1.00010d-00		1.98862d-05	1.50533d-06	9.53992d-05
	7.14351d-07		-1.04162d-04	7.88474d-06	7.14351d-07
1.76715d-00	9.80899d-01		2.57556d-05	-1.94963d-06	1.14035d-04
	-1.95115d-01		-8.07955d-05	6.11599d-06	-1.51278d-05
1.96350d-00	9.24004d-01		9.51950d-06	7.20600d-07	1.24373d-04
	-3.82735d-01		-1.17583d-04	8.90074d-06	-5.14288d-05
2.15985d-00	8.31603d-01		-3.73143d-05	2.82459d-06	1.33015d-04
	-5.55656d-01		-9.41709d-05	7.12848d-06	-8.62367d-05
2.35620d-00	7.07233d-01		-2.36383d-05	1.78936d-06	1.26256d-04
	-7.07234d-01		-8.46694d-05	6.40923d-06	-1.26969d-04
2.55254d-00	5.55682d-01		-6.02440d-05	4.56030d-06	7.11507d-04
	-8.31633d-01		-1.00363d-04	7.59720d-06	-1.63104d-04

Table 5.6a continued

2.74889d-00	3.82772d-01	-7.97571d-05	6.03739d-06	8.80207d-05
	-9.24083d-01	-5.78030d-05	4.37553d-06	-2.03138d-04
2.94524d-00	1.95139d-01	-6.90427d-05	5.22634d-06	4.84728d-05
	-9.81022d-01	-5.97048d-05	4.51949d-06	-2.36557d-04
3.14159d-00	4.08070d-06	-1.05860d-04	8.01335d-06	4.08070d-06
	-1.00026d-00	-4.66466d-05	3.53102d-06	-2.60575d-04

Number of function evaluations = 17

Table 5.6b.

P(EC)E Mode		Milne's Error Estimates		
x	y	Predictor	Corrector	Exact Error
7.85398d-01	7.07116d-01	9.59750d-05	-7.26504d-06	9.38799d-06
	7.07110d-01	-3.22605d-05	2.44203d-06	3.01723d-06
9.81748d-01	8.31489d-01	8.69284d-05	-6.58024d-06	1.95643d-05
	5.55572d-01	-4.85296d-05	3.67355d-06	1.89447d-06
1.17810d-00	9.23909d-01	7.68815d-05	-5.81971d-06	2.93884d-05
	3.82680d-01	-6.63296d-05	5.02097d-06	-2.96380d-06
1.37445d-00	9.80823d-01	6.19845d-05	-4.69206d-06	3.76512d-05
	1.95079d-01	-7.94829d-05	6.01663d-06	-1.15811d-05
1.57080d-00	1.00004d-00	4.53549d-05	-3.43324d-06	4.33187d-05
	-2.34162d-05	-9.00655d-05	6.81771d-06	-2.34162d-05
1.76715d-00	9.80830d-01	2.68900d-05	-2.03549d-06	4.54825d-05
	-1.95128d-01	-0.71742d-05	7.35582d-06	-3.77287d-05
1.96350d-00	9.23923d-01	7.42012d-06	-5.61683d-07	4.34360d-05
	-3.82737d-01	-1.00555d-04	7.61176d-06	-5.35425d-05
2.15984d-00	8.31506d-01	-1.23414d-05	9.34206d-07	3.67293d-05
	-5.55640d-01	-1.00072d-04	7.57516d-06	-6.97133d-05
2.35619d-00	7.07132d-01	-3.16278d-05	2.39413d-06	2.52065d-05
	-7.07192d-01	-9.57419d-05	7.24740d-06	-8.49896d-05
2.55254d-00	5.55579d-01	-4.96993d-05	3.76209d-06	9.02771d-06
	-8.31568d-01	-8.77326d-05	6.64111d-06	-9.80859d-05
2.74889d-00	3.82672d-01	-6.58610d-05	4.98549d-06	-1.13265d-05
	-9.23987d-01	-7.63513d-05	5.77958d-06	-1.07759d-04
2.94524d-00	1.95055d-01	-7.94919d-05	6.01731d-06	-3.50677d-05
	-9.80898d-01	-6.20356d-05	4.69592d-06	-1.12883d-04
3.14159d-00	-6.11308d-05	-9.00680d-05	6.81790d-06	-6.11308d-05
	-1.00011d-00	-4.53354d-05	3.43177d-06	-1.12523d-04

Number of function evaluations = 30

Table 5.7a. Predictor-Corrector Formulas (5.4.12) and (5.4.13)
Linear Multistep Method of Order Four (cf. Fatunla 1987b)
Number of Differential Equations = 2
Predictor Error Constant = 3.375000000d-01
Corrector Error Constant = -3.750000000d-02

P(EC)	Mode
Uniform Stepsize =	1.963495408d-01

Starting Values

x	y
0.000000000d-01	0.000000000d-01
	1.000000000d-00
1.963495408d-01	1.950903220d-01
	9.807852804d-01
3.926990817d-01	3.826834324d-01
	9.238795325d-01
5.890486225d-01	5.555702330d-01
	8.314696123d-01

Numerical Solution		Milne's Error Estimates		
x	y	Predictor	Corrector	Exact Error
7.85398d-01	7.07119d-01	9.28339d-05	-1.03149d-05	1.22872d-05
	7.07108d-01	-3.14305d-05	3.49337d-06	1.55839d-06
9.81747d-01	8.31493d-01	7.83517d-05	-8.70574d-06	2.38618d-05
	5.55588d-01	-7.42684d-05	8.25205d-06	1.78176d-05
1.17809d-00	9.23923d-01	6.24763d-05	-6.94181d-06	4.30912d-05
	3.82711d-01	-4.80812d-05	5.34235d-06	2.75533d-05
1.37444d-00	9.80854d-01	7.07813d-05	-7.86459d-06	6.86558d-05
	1.95109d-01	-8.66469d-05	9.62743d-06	1.91581d-05
1.57079d-00	1.00009d-00	2.30736d-05	-2.56374d-06	8.69089d-05
	1.55637d-05	-9.73872d-05	1.08208d-05	1.55637d-05
1.76714d-00	9.80895d-01	2.36361d-05	-2.62623d-06	1.09926d-04
	-1.95089d-01	-8.41069d-05	9.34522d-06	1.16977d-06
1.96349d-00	9.24010d-01	4.81143d-06	-5.34603d-07	1.30899d-04
	-3.82716d-01	-1.09275d-04	1.21416d-05	-3.22002d-05
2.15984d-00	8.31606d-01	-2.74766d-05	3.05295d-06	1.36583d-04
	-5.55628d-01	-9.01413d-05	1.00157d-05	-5.78847d-05
2.35619d-00	7.07248d-01	-2.95768d-05	3.28631d-06	1.41347d-04
	-7.07198d-01	-8.94682d-05	9.94091d-06	-9.20667d-05
2.55254d-00	5.55709d-01	-5.97471d-05	6.63856d-06	1.38345d-04
	-8.31610d-01	-8.74705d-05	9.71894d-06	-1.40331d-04

Table 5.7a continued

2.74889d-00	3.82799d-01	-6.72450d-05	7.47166d-06	1.15478d-04
	-9.24054d-01	-6.08278d-05	6.75865d-06	-1.74805d-04
2.94524d-00	1.95179d-01	-7.86220d-05	8.73578d-06	8.86600d-05
	-9.80996d-01	-6.19189d-05	6.87987d-06	-2.10222d-04
3.14159d-00	5.41337d-05	-9.77618d-05	1.08624d-05	5.41338d-05
	-1.00025d-00	-3.44792d-05	3.83102d-06	-2.51443d-04

Number of function evaluations = 17

Table 5.7b.

P(EC)E Mode			Milne's Error Estimates	
x	y	Predictor	Corrector	Exact Error
7.85398d-01	7.07119d-01	9.28339d-05	-1.03149d-05	1.22872d-05
	7.07108d-01	-3.14305d-05	3.49227d-06	1.55839d-06
9.81748d-02	8.31482d-01	8.40856d-05	-9.34284d-06	1.22330d-05
	5.55567d-01	-4.57338d-05	5.08153d-06	-3.64195d-06
1.17810d-00	9.23891d-01	7.50865d-05	-8.34294d-06	1.12236d-05
	3.82675d-01	-6.60596d-05	7.33996d-06	-8.51510d-06
1.37445d-00	9.80806d-01	5.97219d-05	-6.63577d-06	2.03534d-05
	1.95080d-01	-7.69559d-05	8.55065d-06	-1.07414d-05
1.57080d-00	1.00002d-00	4.29383d-05	-4.77092d-06	1.63785d-05
	-1.96286d-05	-8.49792d-05	9.44214d-06	-1.96286d-05
1.76715d-00	9.80796d-01	2.73503d-05	-3.03892d-06	1.08548d-05
	-1.95117d-01	-9.62778d-05	1.06975d-05	-2.68510d-05
1.96350d-00	9.23894d-01	6.79935d-06	-7.55483d-07	1.40100d-05
	-3.82713d-01	-9.72773d-05	1.08086d-05	-2.95650d-05
2.15984d-00	8.31474d-01	-1.29843d-05	1.44270d-06	4.26279d-06
	-5.55608d-01	-9.48337d-05	1.05371d-05	-3.81209d-05
2.35619d-00	7.07100d-01	-2.91691d-05	3.24101d-06	-6.63168d-06
	-7.07150d-01	-9.47733d-05	1.05304d-05	-4.31857d-05
2.55254d-00	5.55561d-01	-4.85490d-05	5.39434d-06	-9.17249d-06
	-8.31511d-01	-8.47979d-05	9.42199d-06	-4.17852d-05
2.74889d-00	3.82660d-01	-6.48589d-05	7.20655d-06	-2.31821d-05
	-9.23925d-01	-7.20341d-05	8.00379d-06	-4.54493d-05
2.94524d-00	1.95053d-01	-7.53766d-05	8.37518d-06	-3.68580d-05
	-9.80829d-01	-6.20172d-05	6.89080d-06	1-4.41352d-05
3.14159d-00	-4.14664d-05	-8.76840d-05	9.74267d-06	-4.14664d-05
	-1.00004d-00	-4.37046d-05	4.85606d-06	-3.51279d-05

Number of function evaluations = 30

$$c_1 = \sum_{j=0}^{k} (j - \nu)\alpha_j - \sum_{j=0}^{k} \beta_j \, ,$$

(5.5.4)

$$c_r = \frac{1}{r!}\sum_{j=0}^{k}(j-v)^r\alpha_j - \frac{1}{(r-1)!}\sum_{j=0}^{k}(j-v)^{r-1}\beta_j \ , \ r \geq 2 \ .$$

Normally, v is chosen to minimize the labor of computation in (5.5.4).

Definition 5.1. The LMM (5.5.1) is said to possess an order of accuracy p, provided the coefficients of its associated linear operator $L[y(x), h]$ satisfy the following condition:

$$c_r = 0 \ , \ r \leq p \text{ and } c_{p+1} \neq 0 \ . \tag{5.5.5}$$

c_{p+1} is called the error constant, and the above definition implies that the local truncation error t_{n+k} is

$$t_{n+k} = c_{p+1}h^{p+1}y^{(p+1)}(x_{n+v}) + O(h^{p+2}) \ , \tag{5.5.6}$$

where c_{p+1} turns out to be independent of the choice of v. The first term in (5.5.6) is called the principal local truncation error, and, for sufficiently small h, it often provides a means to control the global accuracy.

It is not necessarily true that an implicit LMM must have error order $p = k+1$, as formulated in Section 5.3. This is shown by the implicit linear multistep method characterized by the polynomials $\rho(r)$ and $\sigma(r)$

$$\sigma(r) = r(r^3-1) \ ,$$

$$\tag{5.5.7}$$

$$\rho(r) = \frac{27}{16}r^4 - \frac{33}{8}r^3 + 9r^2 - \frac{39}{8}r + \frac{21}{16} \ ,$$

with error order $p = 4$, error constant $c_5 = -27/20$.

The question now arises whether the local truncation error $L[y(x_n), h]$ given by (5.5.6) can always be bounded, or, alternatively, can be expressed in the form of the Remainder Theorem

$$L[y(x_n), h] = c_{p+1}h^{p+1}y^{(p+1)}(x_n+\theta h) \ , \tag{5.5.8}$$

for some $\theta \ \varepsilon[0, k]$. In case x_n is a vector, (5.5.8) still holds for some vector θ. This is always possible by appealing to the generalized Remainder Theorem 1.6, whereby the function $y(x_{n+j})$ is expressed as

$$y(x_{n+j}) = \sum_{r=0}^{p}\frac{(jh)^r}{r!}y^{(r)}(x_n) + \frac{1}{p!}\int_{0}^{jh}(jh-s)^p y^{(p+1)}(x_n+s)ds \ , \tag{5.5.9}$$

and its derivative $y\prime(x_{n+j})$ is expressed as

$$y\prime(x_{n+j}) = \sum_{r=0}^{p-1}\frac{(jh)^r}{r!}y^{(r+1)}(x_n)$$

$$\tag{5.5.10}$$

$$+ \frac{1}{(p-1)!} \int_0^{jh} (jh-s)^{p-1} y^{(p+1)}(x_n+s)ds .$$

The adoption of (5.5.9) and (5.5.10) in (5.5.2) with $s = th$ reduces the local truncation error to

$$L[y(x_n), h] = \frac{1}{p!} h^{p+1} \sum_{j=0}^{k} \int_0^j [\alpha_j(j-t)^p$$

(5.5.11)

$$- p\beta_j(j-t)^{p-1}] y^{(p+1)}(x_n+th)dt .$$

Let z_+ denote a non-negative function

$$z_+ = \begin{cases} z, & \text{if } z \geq 0, \\ 0, & \text{otherwise}. \end{cases}$$

(5.5.12)

Using (5.5.12), we have the following identity:

$$\int_0^k (j-t)_+^p y^{(p+1)}(x_n+th)dt = \int_0^j (j-t)_+^p y^{(p+1)}(x_n+th)dt$$

(5.5.13)

$$+ \int_j^k (j-t)_+^p y^{(p+1)} (x_n+th)dt ,$$

whose second term vanishes.

Thus,

$$\int_0^j (j-t)^{p-1} y^{(p+1)}(x_n+th)dt = \int_0^k (j-t)_+^{p-1} y^{(p+1)}(x_n+th)dt .$$ (5.5.14)

Equations (5.5.11), (5.5.13), and (5.5.14) combine to give the following useful expression for the l.t.e.:

$$L[y(x_n), h] = \frac{1}{p!} h^{p+1} \int_0^k F(t) y^{(p+1)}(x_n+th)dt ,$$ (5.5.15)

with the influence function $F(t)$ defined as

$$F(t) = \sum_{j=0}^{k} [\alpha_j(j-t)_+^p - p\beta_j(j-t)_+^{p-1}] .$$ (5.5.16)

In the event that the influence function $F(t)$ is of constant sign in the interval

[0, k], we can safely appeal to the Generalized Mean Value Theorem (Theorem 1.6) to get

$$L[y(x_n), h] = \frac{1}{p!} h^{p+1} y^{(p+1)}(x_n + \theta h) \int_0^k F(t)dt \qquad (5.5.17)$$

for $\theta \, \varepsilon[0, k]$.

Equation (5.5.17) is the Remainder Theorem (Theorem 1.6) approximation of the l.t.e. given by (5.5.8); the error constant c_{p+1} is

$$c_{p+1} = \frac{1}{p!} \int_0^k F(t)dt . \qquad (5.5.18)$$

Example. Obtain the influence function $F(t)$ and, hence, the error constant for Simpson's rule

$$\rho(r) = r^2 - 1 , \sigma(r) = \frac{1}{3}(r^2 + 4r + 1) . \qquad (5.5.19)$$

$$k = 2, \; \alpha_2 = -\alpha_0 = 1 , \; \alpha_1 = 0 ,$$

$$\beta_2 = \beta_0 = \frac{1}{3} , \beta_1 = \frac{4}{3} \; ; p = 4 .$$

For $t \, \varepsilon[0, 1]$,

$$F_1(t) = (-1)(0-t)_+^4 - 4(\frac{1}{3})(0-t)_+^3$$

$$+ (0)(1-t)_+^4 - 4(\frac{4}{3})(1-t)_+^3$$

$$+ (1)(2-t)_+^4 - 4(\frac{1}{3})(2-t)_+^3$$

$$= (t - \frac{4}{3})t^3 \le 0$$

in [0, 1].

Similarly, in the interval [1,2],

$$F_2(t) = (1)(2-t)_+^4 - 4(\frac{1}{3})(2-t)_+^3 = (\frac{2}{3} - t)(2 - t)^3 \le 0 .$$

Hence, $F(t)$ has a constant sign in [0, 2].

The error constant for Simpson's rule (5.5.19) is, therefore, obtained as

$$c_5 = \frac{1}{4!} (\int_0^1 (t^4 - \frac{4}{3}t^3)dt + \int_1^2 (t^4 - \frac{20}{3}t^3 + 16t^2 - 16t + \frac{16}{3})dt) = -\frac{1}{90} .$$

Using (5.5.4) with $v = 1$, the error constant can be obtained as

$$c_5 = \frac{1}{5!}[(2-1)^5 \cdot 1 + (0-1)^5 \cdot - 1]$$

$$- \frac{1}{4!}[(2-1) \cdot \frac{1}{3} + (0-1)^4 \cdot \frac{1}{3}]$$

$$= \frac{1}{60} - \frac{1}{36} = -\frac{1}{90} .$$

Exercise. Derive the influence function and the error constant for the LMM characterized by

$$\rho(r) = r^3 - 1 , \sigma(r) = \frac{3}{8}r^3 + \frac{9}{8}r^2 + \frac{9}{8}r + \frac{3}{8} . \qquad (5.5.20)$$

If, however, the influence function $F(t)$ is not of constant sign in all of the intervals $x_{n+j} \leq x \leq x_{n+j+1}$, $j = 0(1)k-1$, the error constant c_{p+1} can still be bounded in absolute value as

$$|c_{p+1}| \leq \frac{1}{p!}\int_0^k |F(t)| dt . \qquad (5.5.21)$$

The question now arises as to whether the global error $e_{n+k} = y_{n+k} - y(x_{n+k})$ bears any relationship with the l.t.e. The l.t.e. t_{n+k} has been defined previously as

$$t_r = y(x_r) - s_r(h) , \ 0 \leq r \leq k-1$$

$$(5.5.22)$$

$$\sum_{j=0}^k \alpha_j y(x_{n+j}) = h\sum_{j=0}^k \beta_j f(x_{n+j}, y(x_{n+j})) + t_{n+k} , \ 0 \leq j \leq N-k ,$$

where $y(x)$ is the theoretical solution to the IVP.

Subtracting (5.5.22) from (5.5.1) and applying the Mean Value Theorem implies that the global error can be bounded as follows:

$$|e_{n+k}| \leq (1-h|\beta_k|L)^{-1} [\sum_{j=0}^{k-1}[|-\alpha_j|$$

$$(5.5.23)$$

$$+ hL|\beta_j|]|e_{n+j}| , + |t_{n+k}|]$$

where L is the Lipschitz constant of $f(x, y)$ w.r.t. y.

With $n = 0$ and starting errors e_j, $j = 0(1)k-1$, the error e_k depends on the starting errors e_j, $j = 0(1)k-1$ and the l.t.e. t_k. The error e_{k+1} depends on the l.t.e. t_{k+1} and how the errors $e_1, e_2,..., e_k$ grow. In general, the global error e_{n+k} is determined by the l.t.e. t_{n+k} and the mode of propagation of the previous global errors $e_n, e_{n+1},..., e_{n+k-1}$.

If roundoff errors are incorporated, Equation (5.5.23) is replaced by

$$|\tilde{e}_{n+k}| \leq (1-h|\beta_k|L)^{-1}[\sum_{j=0}^{k-1}[|-\alpha_j|$$

$$+ hL|\beta_j|]|e_{n+j}| + |S_{n+k}|] ,$$

where $S_{n+k} = R_{n+k} - t_{n+k}$, $\bar{e}_{n+k} = \bar{y}_{n+k} - y(x_{n+k})$, and R_{n+k} is the roundoff error.

Definition 5.2. The LMM (5.5.1) is said to be consistent provided its error order p satisfies $p \geq 1$. It can be shown that this implies that the first and second characteristic polynomials fulfill

$$\rho(1) = 0 , \quad \rho\prime(1) = \sigma(1) . \tag{5.5.25}$$

The concept of consistency of LMM is very important in the sense that it controls the magnitude of the l.t.e. committed at every integration step, and, hence, is crucial to the convergence of the LMM.

We give the following alternative definition:

Definition 5.3. The LMM (5.5.1) is said to be consistent if

$$\max_{k \leq n \leq N} \frac{t_n}{h} \to 0 \text{ as } h \to 0 ,$$

provided $t_n \neq hy\prime$.

Definition 5.4. The LMM (5.5.1) is convergent if, when it is applied to (2.1.4),

$$\lim_{n \to \infty} y_n = y(x) , \quad h = (x-a)/n \tag{5.5.26}$$

for any arbitrary point $x \in [a, b]$.

Theorem 5.1. *A convergent linear multistep method is consistent.*

Proof. We assume that the LMM (5.5.1) is convergent. This implies that

$$y_n = y(x_n) + r_n(h) ,$$

where $|r_n(h)| \leq r(h)$, $n = 0(1)N$, and $r(h) \to 0$ as $h \to 0$.

Consider first the initial value problem $y\prime = 0$, $y(0) = 1$. We get

$$0 = \sum_{j=0}^{n} \alpha_j y_{n+j} = \sum_{j=0}^{n} \alpha_j + \sum_{j=0}^{n} \alpha_j r_{n+j}(h) ,$$

and so

$$\left| \sum_{j=0}^{n} \alpha_j \right| \le \sum_{j=0}^{n} |\alpha_j| r(h) .$$

Since the LMM is convergent for this problem, $r(h) \to 0$, and we conclude that

$$\sum_{j=0}^{k} \alpha_j = 0 = \rho(1) . \qquad (5.5.27)$$

Consider now the problem $y\prime = 1$, $y(0) = 0$. We get

$$h \sum_{j=0}^{n} \beta_j = \sum_{j=0}^{n} \alpha_j y_{n+j} = \sum_{j=0}^{k} \alpha_j y(x_{n+j}) + \sum_{j=0}^{n} \alpha_j r_{n+j}(h) .$$

Using Equation (5.5.27) yields

$$h \sum_{j=0}^{k} (j\alpha_j - \beta_j) = - \sum_{j=0}^{n} \alpha_j r_{n+j}(h) .$$

Summing n from k to N gives

$$(b - a - (k-1)h) \sum_{j=0}^{k} (j\alpha_j - \beta_j) = - \sum_{n=k}^{N} \sum_{j=0}^{n} \alpha_j r_{n+j}(h)$$

$$= \sum_{n=0}^{k-1} (\sum_{j=0}^{n} \alpha_j) r_n(h) + \sum_{n=k}^{N-k} (\sum_{j=0}^{k} \alpha_j) r_n(h)$$

$$+ \sum_{n=N-k+1}^{N} (\sum_{j=n+k-N}^{k} \alpha_j) r_n(h) .$$

The second term (due to (5.5.27)) vanishes identically, while the first and third terms are bounded by some constant times $r(h)$. Hence,

$$\sum_{j=0}^{k} (j\alpha_j - \beta_j) = 0 = \rho\prime(1) - \sigma(1) . \qquad (5.5.28)$$

                                                    ■

A normalized LMM (5.5.1) with stepnumber $k(\alpha_k = 1)$ has $2k$ or $2k+1$ unknown coefficients $\{\alpha_j, \beta_j \mid j = 0(1)k\}$, depending on whether it is explicit or implicit. The numerical values of these coefficients are usually determined by ensuring that sufficiently many coefficients c_r of (5.5.4) vanish; i.e.,

$$c_r = 0 \begin{cases} r = 0(1)2k-1 , & \beta_k = 0 \text{ explicit} , \\ r = 0(1)2k , & \beta_k \neq 0 \text{ implicit} . \end{cases}$$

The resultant integration formula is then of highest order:

$$p = \begin{cases} 2k-1, & \text{for explicit scheme}, \\ 2k, & \text{for implicit scheme}. \end{cases} \tag{5.5.29}$$

Such formulas are classified as being optimal.

Ideally, one would wish for a k-step LMM of the highest possible order, but these goals are not always attainable due to the Dahlquist barrier theorems (1959, 1963). Optimal formulas are now only of historical interest because their poor stability property precludes their general use.

Definition 5.5. The LMM (5.5.1) is said to be zero-stable if no root of the first characteristic polynomial $\rho(r)$ has modulus greater than one, and if every root with unit modulus is simple.

The main consequence of zero-stability is to control the propagation of the errors as the integration progresses.

Dahlquist (1956, 1959) established the following important theorems which guide numerical analysts in the formulation of new integration formulas:

Theorem 5.2. *The LMM (5.5.1) is convergent* **iff** *it is consistent and zero-stable.*

Theorem 5.3. *A zero-stable LMM (5.5.1) is, at best, of order $p = k+1$ for k odd; and of order $p = k+2$ for k even.*

The proofs of these theorems are available in the original papers of Dahlquist (1956) and Henrici's treatise of (1962).

Definition 5.6. A zero-stable LMM with stepnumber k is optimal if its order $p = k+2$.

Optimal LMMs are such that all the roots of the first characteristic polynomials $\rho(r)$ lie on the unit circle.

An example of optimal LMM is Simpson's rule, which is given by

$$\rho_2(r) = r^2 - 1, \; \sigma_2(r) = \frac{1}{3}(r^2 + 4r + 1). \tag{5.5.30}$$

Equation (5.5.30) is called weakly stable because its first characteristic polynomial has more than one root on the unit circle. It is unstable when $\partial f / \partial y < 0$, but Stetter (1965) pointed out that by incorporating a predictor characterized by

$$\rho_1(r) = r^2 + 4r - 5, \; \sigma_1(r) = 4r + 2 \tag{5.5.31}$$

into (5.5.30), the resultant PECE algorithm has a non-vanishing interval of stability $(-1, 0)$.

Example. Generate a numerical solution to the IVP $y' = -10y$, $y(0) = 1$, $0 \le x \le 2$, using stepsizes $\{h = 2^{-q}, q = 2(1)7\}$ with Simpson's rule:

Table 5.8. Simpson's rule in PECE Mode
$y' = -10y$, $y(0) = 1$, $0 \leq x \leq 2$, $h = 2^{-q}$; Exact Solution: 2.06115362(-9)

q	(5.5.33) (5.5.31)	(5.5.32) (5.5.31)
2	-4.55685768(0)	2.88884241(5)
3	6.40071163(-3)	3.75526455(2)
4	-2.92759404(-3)	2.72822315(-9)
5	-2.254388263(-2)	2.07530637(-9)
6	-1.08982166(-2)	2.06182838(-9)
7	-1.30953319(-3)	2.06119136(-9)

(i) $\rho_1(r) = r^4 - r$, $\rho_1(r) = \dfrac{1}{8}(21r^3 - 9r^2 + 15r - 3)$, and   (5.5.32)

(ii)   Stetter's predictor (5.5.31).

Table 5.8 clearly illustrates the stabilizing effect of Stetter's predictor.

The integration formula can behave violently in the event that it is zero-unstable, i.e., if any of the roots of the first characteristic polynomial $\rho(r)$ falls outside the unit circle. This point is well made with the numerical solution of the IVP $y' = 3x^2$, $y(0) = 1$ with the LMM

$$\rho(r) = r^2 + 2r - 3 ,$$

(5.5.33)

$$\sigma(r) = 3r + 1 ,$$

whose order is two, but with the zeros of $\rho(r)$ located at +1 and -3.

As can be observed from Table 5.9, the zero-unstable scheme "explodes" with decreasing mesh size.

We now consider the application of the LMM (5.5.1) to the scalar test problem $y' = \lambda y$ to obtain the following difference equation of order k:

$$(\rho(E) - z\sigma(E))y_n = 0 , z = \lambda h ,$$ (5.5.34)

where E is the shift operator specified by (1.1), with $\rho(E)$ and $\sigma(E)$ being the first and second characteristic polynomials of the integration formula, respectively.

The l.t.e. t_{n+k} of (5.5.1) satisfies

$$(\rho(E) - z\sigma(E))y(x_n) = t_{n+k} ,$$ (5.5.35)

where $y(x)$ is the theoretical solution to the IVP.

By considering the effect of the roundoff errors r_{n+k}, (5.5.34) is replaced by

Table 5.9. Effect of Zero-Instability of LMM (5.5.34)
$y\prime = 3x^2$, $y(0) = 1$; Exact Solution: $y(x) = x^3$

x	exact	$y(x, 1/4)$	$y(x, 1/8)$	$y(x, 1/16)$
0.25	0.0156250	0.0156250	0.013672	0.014160
0.50	0.1250000	0.109375	0.113281	0.024414
1.00	1.0000000	0.906250	0.195312	-6.558413×10^2
1.50	3.3750000	2.640625	-61.503906	-4.309529×10^6
2.00	8.0000000	1.562500	-5.246734×10^3	-2.827484×10^{10}
2.50	15.6250000	-42.078125	-4.256172×10^5	-1.855112×10^{14}
3.00	27.0000000	-492.031250	-3.447623×10^7	-1.217139×10^{18}

$$(\rho(E) - z\sigma(E))\tilde{y}_n = r_{n+k} , \qquad (5.5.36)$$

where $\tilde{y}_n$ is the numerical solution which incorporates the effect of roundoff errors.

Subtracting (5.5.35) from (5.5.36) implies

$$(\rho(E) - z\sigma(E))e_n = s_{n+k} . \qquad (5.5.37)$$

By making the simplifying assumption that $s_{n+k} = s$, a constant, the stability equation becomes

$$(\rho(E) - z\sigma(E))e_n = s . \qquad (5.5.38)$$

The general solution of this problem can be written as

$$e_n = \sum_{r=0}^{k} \gamma_r \xi_r^n + P_n , \qquad (5.5.39)$$

where P_n is a particular solution specified as

$$P_n = \frac{-s}{z\sigma(1)} , \qquad (5.5.40)$$

and ξ_r, $r = 1(1)k$ are the roots of the stability polynomial $\rho(r) - z\sigma(r) = 0$. The constant coefficients γ_r, $r = 1(1)k$ are uniquely specified if the initial errors $e_0, e_1, ..., e_{k-1}$ are known.

The difference equation (5.5.34) (whose characteristic equation has distinct roots) also has the solution

$$y_n = \sum_{r=1}^{k} \delta_r \xi_r^n , \qquad (5.5.41)$$

whose coefficients δ_r, $r = 1(1)k$ can be obtained from the initial condition y_0 and the necessary additional starting values $y_1, y_2, ..., y_{k-1}$.

The IVP under consideration has a theoretical solution

$$y(x_n) = (e^z)^n y_0 , \tag{5.5.42}$$

and, hence, there is at least one root ξ_r (ξ_1, say) such that $\xi_1 = e^z + O(z^{p+1})$. The other $k-1$ roots $\xi_2, \xi_3, ..., \xi_k$ are spurious roots, which emanate from a replacement of a first order IVP by a k-th order difference equation. It is, then, only desirable that the contributions of these extra parasitic roots to the numerical solution y_n are driven to zero with increasing n. This is possible provided that $\xi_r^n \to 0$, $r = 2(1)k$, and this holds if $|\xi_r| < 1$ for $r = 2(1)k$.

Definition 5.7. The LMM (5.5.1) is said to be strongly stable if $|\xi_r| < 1$, $r = 2(1)k$ for $z = 0$.

Definition 5.8. The LMM (5.5.1) is said to be weakly stable if there is at least one ξ_r, $r = 2(1)k$ such that $|\xi_r(0)| = 1$ and $|\xi_r(z)| > 1$ for $\text{Re}(z) < 0$.

Definition 5.9. The LMM (5.5.1) is said to be absolutely stable for a given z if $|\xi_r| \le 1$, $r = 1(1)k$, and the region of absolute stability (RAS) is the set $l = \{z \varepsilon C : |\xi_r| < 1 , r = 1(1)k\}$.

Definition 5.10. The LMM is said to be relatively stable for a given z if $|\xi_r| < |\xi_1|$, $r = 2(1)k$.

If the parameter $z = \lambda h$ approaches zero, then the stability polynomial

$$\Pi(r, z) = \rho(r) - z\sigma(r) \tag{5.5.43}$$

approaches $\rho(r)$, and, hence, the stability of (5.5.1) is determined by the location of the roots of $\rho(r)$. This gives birth to the concept of zero-stability. In the opposite, case when $z \to \infty$, the stability of (5.5.1) is determined by the position of the roots of $\sigma(r)$, and this explains why the backward differentiation formulas (5.3.18), whose second characteristic polynomials are $\sigma(r) = r^k$, are well suited to stiff initial value problems: all the roots of $\sigma(r)$ are located at the origin, with the result that such formulas are highly stable.

It is, in general, impossible to identify ξ_1 away from the origin; $\xi_1, ..., \xi_k$ "live" on a Riemann surface and cannot be identified beyond branch points.

Table 5.10. Existing Adams Codes

Author	Year	Code	Mode	Type	Stepsize Strategy	Local Extrapolation	Stiff Option
*Krogh	1970	DVDQ	Scaled Divided Difference	PECE	Variable Coefficient	Yes	No
*Gear	1971a	DIFSUB	Nordsieck	AM	Fixed Coefficient	No	Yes
Sedgwick	1973	VOAS	Divided Difference	PECE	Variable Coefficient	No	No
*Hindmarsh	1972	GEAR	Nordsieck	AM	Fixed Coefficient	No	Yes
Byrne and Hindmarsh	1975	EPISODE	Nordsieck	AM	Variable Coefficient	No	Yes

Table 5.10. Existing Adams Codes (continued)

Author	Year	Code	Mode	Type	Stepsize Strategy	Local Extrapolation	Stiff Option
+Shampine and Gordon	1975	STEP	Scaled Divided Difference	PECE	Variable Coefficient	Yes	No
Hindmarsh	1980	LSODE	Nordsieck	AM	Fixed Coefficient	No	Yes
Shampine and Watts	1979	DEPAC	Scaled Divided Difference	PECE	Variable Coefficient	Yes	Yes
Gear	1980	-	Modified Divided Difference	AM	Variable Coefficient	No	Yes
Petzold	1983	LSODA	Nordsieck	AM	Fixed Coefficient	No	Yes

*Only of historical interest
+Incorporated into DEPAC

5.6. Automatic Implementation of the Adams Scheme

The currently available codes based on the Adams formula are listed in Table 5.10.

Most modern implementations of the Adams code adopt variable order p in the range $2 \le p \le 13$, starting off with an order two scheme (i.e., stepnumber $k = 1$), since, at the start of integration, only one value of the dependent variable is available.

The k-step Adams-Bashforth scheme (predictor formula) was obtained in Section 5.2 from the identity

$$y(x) = y(x_n) + \int_{x_n}^{x} f(t, y(t))dt . \tag{5.6.1}$$

To advance the integration from x_n to x_{n+1} (i.e., $h_{n+1} = x_{n+1} - x_n$), the integrand is approximated by a polynomial $Q_{k-1,n}(t)$ of degree $k-1$ interpolating the available computed values f_{n+1-j}, $j = 1(1)k$. This yields the k-th order Adams-Bashforth scheme

$$y_{n+1}^p = y_n + \int_{x_n}^{x_{n+1}} Q_{k-1,n}(t)dt . \tag{5.6.2}$$

The equivalent Adams-Moulton scheme is obtained by replacing the integrand with a polynomial $P_{k-1}(t)$ of degree $k-1$, but now interpolating at the past derivatives f_{n+1-j}, $j = 0(1)k-1$,

$$y_{n+1} = y_n + \int_{x_n}^{x_{n+1}} P_{k-1,n}(t)dt . \tag{5.6.3}$$

Since $P_{k-1,n}(t)$ and $Q_{k-1,n}(t)$ coincide at $t = x_{n+1-j}$, $j = 1(1)k$ and

$$Q_{k-1,n}(x_{n+1}) = f_{n+1}^p , \quad P_{k-1,n}(x_{n+1}) = f_{n+1} ,$$

Equations (5.6.2) and (5.6.3) yield

$$y_{n+1} = y_{n+1}^p + \int_{x_n}^{x_{n+1}} (P_{k-1,n}(t) - Q_{k-1,n}(t))dt .$$

Adopting the identity

$$\gamma_{k-1,n+1} = \frac{1}{h_{n+1}} \int_{x_n}^{x_{n+1}} \prod_{j=1}^{k} \left(\frac{t - x_{n+1-j}}{x_{n+1} - x_{n+1-j}} \right) dt$$

in the last expression yields the nonlinear equation

$$y_{n+1} = y_{n+1}^p + \gamma_{k-1,n+1} h_{n+1} (f_{n+1} - f_{n+1}^p) ,$$

which is normally solved by the one point iteration scheme as

$$y_{n+1}^{[0]} = y_{n+1}^P ,$$

$$\tag{5.6.4}$$

$$y_{n+1}^{[s+1]} = y_{n+1}^P + \gamma_{k-1,n+1} h_{n+1} (f(x_{n+1}, y_{n+1}^{[s]}) - f_{n+1}^P) , \ s = 0, 1,....$$

Definition 5.11. The type of iteration is called Adams-Moulton (AM) if (5.6.4) is iterated to convergence.

Some codes implement the PECE mode.

An estimate of the local error can be obtained by generating a numerical solution of order $p = k+1$; i.e., replace integrand in (5.6.3) by a polynomial $P_{k,n}(t)$ of degree k to get

$$y_{n+1}^* = y_n + \int_{x_n}^{x_{n+1}} P_{k,n}(t) dt , \tag{5.6.5}$$

and then estimate the local error as

$$e_{n+1} = y_{n+1}^* - y_{n+1} .$$

From (5.6.2) and (5.6.5) we have

$$y_{n+1}^* = y_{n+1}^P + \int_{x_n}^{x_{n+1}} (P_{k,n}(t) - P_{k-1,n}(t)) dt ;$$

i.e.,

$$y_{n+1}^* = y_{n+1}^P + \gamma_{k,n+1} h_{n+1} (f_{n+1} - f_{n+1}^P) . \tag{5.6.6}$$

Subtracting (5.6.4) from (5.6.6) gives the l.e. as

$$e_{n+1} = (\gamma_{k,n+1} - \gamma_{k-1,n+1}) h_{n+1} (f_{n+1} - f_{n+1}^P) . \tag{5.6.7}$$

Definition 5.12. By adopting $y_{n+1}^* = y_{n+1} + e_{n+1}$ as the solution at $x = x_{n+1}$, local extrapolation is said to be performed.

For the non-stiff methods like the Adams scheme, factors which control the choice of stepsize are enumerated by Petzold (1983):

(i) Accuracy over the next step. If ε is the allowable error tolerance and RTOL, ATOL are the local error tolerance parameters, Shampine and Watts (1980) suggested the mixed error control:

$$|e_{n+1}| \leq RTOL * |y_{n+1}| + ATOL \tag{5.6.8}$$

for each component of the dependent variable. The control (5.6.8) reduces to pure relative error if ATOL = 0 or $|y_{n+1}| \gg 0$, while it reduces to pure absolute error on a particular component if RTOL = 0 or $|y_{n+1}| \ll ATOL$. Byrne (1980) suggested an alternative error control strategy

$$|e_{n+1}| \le \varepsilon^* \max(|y_{n+1}|, \text{FLOOR}), \qquad (5.6.9)$$

where FLOOR is the largest value of the dependent variable encountered during the integration process. Controls (5.6.8) and (5.6.9) are equivalent provided RTOL $= \varepsilon$, ATOL $= \varepsilon^*$ FLOOR.

(ii) For the implicit Adams-Moulton scheme the meshsize is chosen as to ensure the rapid convergence of the functional iteration.

(iii) Finally, the meshsize must be chosen as to ensure the stability of the method. A stability constraint compels the adoption of lower order Adams formulas, since the region of absolute stability (RAS) shrinks with increasing stepnumber.

Accuracy considerations suggest the adoption of meshsize

$$h_{n+2} \le \left(\frac{\varepsilon}{e_{n+1}} \right)^{1/p+1} \qquad (5.6.10)$$

over the next integration step. Convergence criterion implies

$$h_{n+2} \gamma \frac{\partial f}{\partial y} \le r = \frac{1}{2}, \qquad (5.6.11)$$

where r is the rate of convergence and $\gamma = \gamma_{k,n+1}$.

Stability constraint also implies

$$h_{n+2} \rho \left(\frac{\partial f}{\partial y} \right) \le r_p, \qquad (5.6.12)$$

where r_p is the radius of the largest disk contained in the RAS of the method. The minimum of the three meshsizes obtained from (5.6.10)—(5.6.12) is adopted as the next integration meshsize.

The implementation of the variable step Adams methods can be effected in two ways,

(i) *Fixed coefficient approach*, whereby (5.6.3) is expressed as

$$y_{n+1} = y_n + h \sum_{j=0}^{k} \beta_j f_{n+1-j}, \qquad (5.6.13)$$

whose constant coefficients β_i, $i = 0(1)k$ are normally precomputed, and $h_{n-j+1} = h$, constant for $j = 0(1)k-1$. A change in stepsize is effected by interpolation at evenly spaced points.

(ii) *Variable coefficient approach*, such that the coefficients β_j in (5.6.13) are dependent on the meshsize of previous steps.

It has been established both theoretically and empirically that the variable coefficients implementation of the Adams methods possesses superior stability characteristics than the fixed coefficients mode (cf. Gear and Tu 1974, Gear and Watanabe 1974, Zlatev 1978, and Piotrowski 1969).

Another striking feature of the implementation of the Adams methods is the mode of storage of the past values of the derivatives. While some authors advocate

(i)   the adoption of storing the past values of the derivatives f_{n+1-j}, $j = 1(1)k-1$, others advocate

(ii) the use of the Nordsieck (1962) array, which is merely the Taylor series representation of the polynomial, while

(iii) the concept of storing the divided difference seems to be gaining ground.

It is possible that a user-specified output point $xout$ is sandwiched in $x_n \leq x \leq x_{n+1}$. The numerical estimate of $y(xout)$ is readily obtained from the identity

$$y = y_n + \int_{x_n}^{xout} P_{k,n}(t)dt , \qquad (5.6.14)$$

since a representation of $P_{k,n}(t)$ is available. Codes based on Adams methods are, thus, well-suited to nonstiff problems where a sizeable quantity of output points are desired (cf. Stetter 1979c).

In practice, the l.t.e.'s are supposed to be about one-tenth the desired global error, and the coefficients of the LMM (5.5.1) are chosen so as to ensure stability. Roundoff errors also creep in, since the mathematical universe of most electronic computers comprises only rational numbers with finite decimal representations. Unfortunately, most numbers have infinite decimal representations.

6. NUMERICAL TREATMENT OF SINGULAR/ DISCONTINUOUS INITIAL VALUE PROBLEMS

6.1. Introduction

The mathematical formulation of physical phenomena in simulation, electrical engineering, control theory, and economics often leads to an initial value problem

$$y\prime = f(x, y) , y(0) = y_0 \qquad (6.1.1)$$

in which there is a pole in the solution or a discontinuous low order derivative.

A simple example is the innocent looking IVP

$$y\prime = y^2 , y(0) = 1 , 0 < x \leq 2 , \qquad (6.1.2)$$

whose theoretical solution

$$y(x) = \frac{1}{1-x} \qquad (6.1.3)$$

has a simple pole at $x = 1$. A similar example was given by (2.3.1a), and such problems are classified as singular IVPs.

Another category of problems arises when the righthand side of the function f in (6.1.11) contains discontinuities in the form of finite jumps in the components of f itself or in some derivatives of f-like problems (2.3.1b).

The switching on and off of electrical circuits and the state of the economy of a nation disrupted by an unforeseen disaster are practical examples of such problems.

In general, there is no clue regarding the location of the singularities. The situation is even worse as $f(x, y)$ is always nonlinear. Such problems can be identified by unbounded Lipschitz constants with respect to the dependent variable y, as illustrated in Section 2.3. Hence, they violate the hypothesis of the existence Theorem 2.1.

Incidentally, the development of the conventional integration formulas (linear multistep methods) discussed in Chapter 5 is based exclusively on polynomial interpolation. Consequently, such schemes perform very poorly in the neighborhood of a singularity. Shampine and Gordon (1975) in the code STEP, Gear and Østerby (1984), Hindmarsh (1979) in the code LSODAR, and Enright et al. (1986) built devices into existing algorithms to cope with discontinuities.

Alternative strategies are based on non-polynomial interpolating functions, either by perturbed polynomials or rational functions. The resultant algorithms often behave nicely in the neighborhood of a singularity, provided the mesh size is chosen as to sandwich the point of singularity. The snags with such schemes are the limited analysis and theoretical backing.

In the event that the points of discontinuities are known beforehand, the conventional algorithms can be efficiently implemented to solve such differential systems by ensuring that such points are made meshpoints.

The theory of ordinary nonlinear differential equations offers no clue as to the location and the nature of singularities in the solution of an equation. Hence, singularities have to be detected heuristically. For instance, the function

$$y(x) = \frac{J_1(x)}{J_0(x)} , \tag{6.1.4}$$

where $J_r(x)$ is the Bessel function of the first kind and a solution to a first order nonlinear differential equation of the Riccati type

$$y\prime = 1 + y^2 - \frac{y}{x} , \; y(0) = 0 , \tag{6.1.5}$$

has an infinite number of poles on the positive real axis. The existing algorithms designed for singular/discontinuous IVPs include

(a) The switching function techniques:
     O'Regan (1970) - Fractional Step, Ellison (1981) - Interpolation, Hay et al. (1974) - Inverse Interpolation, Evans and Fatunla (1975) - Fractional Step, Mannshardt (1978) - Fractional Step, Halin (1976, 1983) - Taylor Series, Carver (1977, 1978) - Discontinuity Condition built in as additional differential equation.

(b) Perturbed polynomial techniques:
     Lambert and Shaw (1966), Shaw (1967).

(c) Rational function methods:
     Lambert and Shaw (1965), Luke et al. (1975), Fatunla (1982a), Niekerk (1987).

(d) Extrapolation process:
     Fatunla (1986).

(e) Gear and Østerby (1984) proposed the use of a local error estimator in an
     automatic code, GEAR, developed by Hindmarsh (1974).

The switching function technique adjoins to (6.1.1) a singularity function

$$\phi(x, y) = 0 . \tag{6.1.6}$$

A change in sign of the switching function (6.1.4) recalls the existence of a
singularity. The occurrence of the relation

$$\phi(x_n, y_n) \, \phi(x_{n+1}, y_{n+1}) < 0 , \tag{6.1.7}$$

thus implies the existence of a singularity in the interval $x_n \leq x \leq x_{n+1}$.

Gear and Østerby (1984) proposed an efficient method (using local error
estimators) to detect and locate the point of discontinuity without using the
singularity function. Provision is also provided to pass the discontinuity and
restart the integration process. Gear (1980d), Ellison (1981), and Enright et al.
(1986) provide Runge-Kutta-like formulas that enable an efficient restart of the
multistep algorithm at discontinuities.

6.2. Non-Polynomial Methods

Lambert and Shaw (1966) proposed that the theoretical solution to (6.1.1) be
represented by either of the following perturbed polynomials

$$F(x) = P_L(x) + \begin{cases} \rho \, |A + x|^N , & N \notin \{0, 1, ..., L\} , \text{ or} \\ \rho \, |A + x|^N \log |A + x| , & N \in \{0, 1, ..., L\} , \end{cases} \tag{6.2.1}$$

with $P_L(x)$ a polynomial of degree L defined by

$$P_L(x) = \sum_{r=0}^{L} a_r x^r , \tag{6.2.2}$$

and the second term on the r.h.s. being the perturbation term.

A and N are the singularity parameters, with A controlling the location of the
singularity, while N determines the nature of the singularity.

The following situations may arise in case of the interpolating function
(6.2.1a):

(i) N real but negative suggests the existence of a singularity; at $x = -A$;

(ii) N integral with $N - L > 0$ suggests the adoption of the conventional linear
      multistep method in an interval that precludes the point $x = -A$;

(iii) N, a positive non-integral number gives a new range of interpolating
      functions.

In the event that $N = 0$ in Equation (6.2.1b), there is a logarithmic singularity
at $x = -A$.

By imposing the constraints

$$F(x_{n+j}) = y(x_{n+j}) , \; j = 0, 1 ,$$

(6.2.3)

$$F^{(s)}(x_n) = y^{(s)}(x_n) , \; s = 0(1)(L + 1)$$

on the interpolating functions (6.2.1a,b), Lambert and Shaw obtained the following one-step integration formulas:

$$y_{n+1} = y_n + \sum_{r=1}^{L} \frac{h^r}{r!} y_n^{(r)} + \frac{(A+x_n)^{L+1}}{\alpha_L^N} y_n^{(L+1)} [(1 + \frac{h}{A+x_n})^N - 1$$

(6.2.4)

$$- \sum_{r=1}^{L} \frac{\alpha_{r-1}^N}{r!} (\frac{h}{A+x_n})^r]$$

from (6.2.1a), and

$$y_{n+1} = y_n + \sum_{r=1}^{L} \frac{h^r}{r!} y_n^{(r)} + (-1)^{L-N} \frac{(A+x_n)^{L+1} y_n^{(L+1)}}{N!(L-N)!} \times$$

(6.2.5)

$$[(1 + \frac{h}{A+x_n})^N \log(1 + \frac{h}{A+x_n}) - \sum_{r=1}^{L} \{ \frac{h^r \alpha_{r-1}^N}{r!(A+x_n)^r} \sum_{j=0}^{r-1} \frac{1}{N-j} \}]$$

from (6.2.1b), where

$$\alpha_r^m = m(m-1)...(m-r) , \; r > 0 .$$ (6.2.6)

The integration formulas (6.2.4) and (6.2.5) are both of order $L+1$, with local truncation error

$$t_{n+1} = \sum_{r=L+2}^{\infty} [y_n^{(r)} - \frac{\alpha_{r-L-2}^{N-L-1}}{(A+x_n)^{r-L-1}} y_n^{(L+1)}] \frac{h^r}{r!} .$$ (6.2.7)

The singularity parameters can be obtained by ensuring that the first two terms in (6.2.7) vanish. This leads to

$$N_{(n)} = L + 1 + \frac{(y_n^{(L+2)})^2}{(y_n^{(L+2)})^2 - y_n^{(L+1)} y_n^{(L+3)}}$$ (6.2.8)

and

$$A_{(n)} = -x_n + \frac{y_n^{(L+2)} y_n^{(L+1)}}{(y_n^{(L+2)})^2 - y_n^{(L+1)} y_n^{(L+3)}} .$$ (6.2.9)

Example 6.1. Solve the initial value problem

$$y' = 1 + y^2 \,,\; y(0) = 1 \,,\; 0 \le x \le 0.75 \,,$$

using a uniform meshsize $h = 0.05$.

Shaw (1967) extended this discussion to multistep methods, thereby eliminating the need to generate the higher derivatives analytically. The singularity parameters are obtained by solving a pair of nonlinear equations.

The unfortunate aspect is that these schemes can only handle IVPs whose singularities are restricted to those of (6.2.1).

Lambert and Shaw (1965) presented an alternative procedure that was based on a local representation of the theoretical solution to (6.1.1) by specialized form of rational function

$$F(x) = \frac{P_m(x)}{(b+x)} \,, \tag{6.2.10}$$

where $P_m(x)$ is a polynomial of degree m.

The resultant integration formulas emanating from (6.2.10) can only cope with special singular initial value problems. Luke et al. (1975) suggested a replacement of the interpolating function (6.2.10) by the generalized rational function

$$F(x) = \frac{P_\mu(x)}{Q_\nu(x)} \,. \tag{6.2.11a}$$

The singularities are specified by the zeros of $Q_\nu(x)$.

In (6.2.11a), the polynomials $P_\mu(x)$ and $Q_\nu(x)$ are given as

$$P_\mu(x) = \sum_{r=0}^{\mu} a_r x^r \tag{6.2.11b}$$

Table 6.1. (Lambert 1973a,b)

x	Theoretical Solution	$A_{(n)}$	$N_{(n)}$	Error $L=1$	Numerical Solution $L=3$
0.00	1.000000000	1.000000	-2.000000	-	-
0.10	1.223048880	0.871052	-1.459539	2×10^{7}	2×16^{-8}
0.20	1.508497647	0.818607	-1.209581	5×10^{-7}	3×10^{-8}
0.30	1.895765123	0.797043	-1.089014	1×10^{-6}	6×10^{-8}
0.40	2.464962757	0.788794	-1.032812	3×10^{-6}	1×10^{-7}
0.50	3.408223442	0.786114	-1.009367	5×10^{-6}	2×10^{-7}
0.60	5.331855223	0.785478	-1.001613	1×10^{-5}	5×10^{-7}
0.65	7.340436575	0.785415	-1.000454	3×10^{-5}	1×10^{-6}
0.70	11.681373800	0.785400	-1.000071	7×10^{-5}	3×10^{-6}
0.75	28.238252850	0.785399	-1.000002	4×10^{-4}	4×10^{-6}

and

$$Q_v(x) = 1 + \sum_{r=1}^{v} b_r x^r .$$ (6.2.11c)

We specify the error function $E_{\mu,v}(x)$ as follows:

$$E_{\mu,v} = Q_v(x)y(x) - P_\mu(x) ,$$ (6.2.12)

which, on differentiating w.r.t. x, gives

$$E'_{\mu,v}(x) = Q_v(x)y'(x) + Q'_v(x)y(x) - P'_\mu(x) .$$ (6.2.13)

The development of the integration algorithm is illustrated with the simple case whereby $\mu = v = 1$ in (6.2.11), which leads to

$$E_{1,1}(x) = (1+b_1 x)y(x) - (a_0 + a_1 x)$$ (6.2.14)

and

$$E'_{1,1}(x) = (1+b_1 x)y'(x) + b_1 y(x) - a_1 .$$ (6.2.15)

Ensuring that the interpolating function coincides with the theoretical solution at x_{n+j}, $j = 0(1)2$ implies

$$E_{1,1}(x_{n+j}) = 0 , \ j = 0(1)2 .$$ (6.2.16a)

We further constrain the interpolating function to satisfy the differential equation at $x = x_{n+1}$. This yields

$$E'_{1,1}(x_{n+1}) = 0 .$$ (6.2.16b)

The derivation of the integration schemes can be greatly simplified by introducing the following transformation in (6.2.16):

$$x = x_0 + th .$$ (6.2.17)

This procedure is equivalent to using

$$x_n = 0 , \ x_{n+j} = j$$ (6.2.18)

and, after the formulas are obtained, replacing y'_j by hy'_j.

The constraints (6.2.16) and (6.2.18) adopted in (6.2.14) and (6.2.15) yield the following set of equations:

$$y_n = a_0 ,$$ (6.2.19a)

$$(1+b_1)y_{n+1} = a_0 + a_1 ,$$ (6.2.19b)

$$(1+2b_1)y_{n+2} = a_0 + 2a_1 ,$$ (6.2.19c)

$$(1+b_1)hy'_{n+1} + b_1 y_{n+1} = a_1 . \tag{6.2.19d}$$

Substitute (6.2.19a) and (6.2.19d) in (6.2.19b) to get

$$b_1 = \frac{y_{n+1} - y_0 - hy'_{n+1}}{hy'_{n+1}} . \tag{6.2.20}$$

Also, adopting (6.2.19b) in (6.2.19c) gives

$$(1+2b_1)y_{n+2} = (1+b_1)y_{n+1} + a_1 .$$

By inserting (6.2.19d) and (6.1.20) into the last equation, we obtain predictor formulas as

$$y_{n+2} = \frac{2y_{n+1}^2 - 2y_n y_{n+1} + hy'_{n+1} y_n}{2y_{n+1} - 2y_n - hy'_{n+1}} . \tag{6.2.21}$$

Example. Solve Example 6.1 adopting the integration formula (6.2.21) in $0 \le x \le 1$ with uniform meshsize $h = 0.05$. Use the exact solution to generate y_1; i.e., $y_1 = \tan(0.05 + \pi/4)$. The details of the numerical experiment are given in Table 6.2.

In general, the relevant equations for the case $\mu = \nu$ are

$$E_{\mu,\nu}(x_{n+j}) = 0 , \; j = 0(1)\nu+1 \tag{6.2.22}$$

and

Table 6.2. Numerical Results of (6.2.21)
$y' = 1 + y^2$, $y(0) = 1$, $0 \le x \le 1 \cdot y(x) = \tan(x + \pi/4)$
Uniform meshsize $h = 0.05$

x	$y(x)$	$y(x, h)$	$y(x, h) - y(x)$	root of $Q_1(x)$
0.10	1.223048880	1.222841398	-0.000207482	-19.370767447
0.20	1.508497647	1.507953126	-0.000544521	-15.934485378
0.30	1.895765123	1.8946198553	-0.001145268	-12.986898396
0.40	2.464962757	2.462611492	-0.002351265	-10.382285069
0.50	3.408223442	3.402986432	-0.005237011	-8.019642767
0.60	5.331855223	5.317214061	-0.014641163	-5.825013860
0.65	7.340436575	7.313159298	-0.027277277	-4.749613200
0.70	11.681373800	11.601918727	-0.079455073	-3.741044037
0.75	28.238252850	27.781116329	-0.457136521	-2.721379543
0.80	-68.479668346	-71.748243078	-3.268574732	-1.720345703
0.85	-15.457896135	-15.619146183	-0.161250047	-0.721182751
0.90	-8.687629546	-8.745226696	-0.057597149	-.279114597
0.95	-6.020299716	-6.048293840	-0.027994123	1.291329417
1.00	-4.588037825	-4.606440346	-0.018402521	2.297629196

$$E'_{\mu,\nu}(x_{n+j}) = 0 , j = 1(1)\nu$$

for the predictor formula; while the corrector formula can be obtained from the linear system

$$E_{\mu,\nu}(x_{n+j}) = 0 , j = 1(1)\nu+1 \tag{6.2.23}$$

and

$$E'_{\mu,\nu}(x_{n+j}) = 0 , j = 1(1)\nu+1 .$$

The system of Equations (6.2.22) is known to have a solution in the unknown coefficients of P and Q if and only if the determinant $|\Delta_p| = 0$ where the predictor matrix Δ_p is given by

$$\Delta_p = \begin{vmatrix} R_{\nu+2,\mu+1} & S_{\nu+2,\mu+1} \\ T_{\nu,\mu+1} & U_{\nu,\mu+1} \end{vmatrix}, \tag{6.2.24}$$

and whose elements are specified as

$$r_{ij} = (i-1)^{j-1} ,$$

$$s_{ij} = (i-1)^{j-1}y_{i-1} ,$$

$$\tag{6.2.25}$$

$$t_{ij} = (j-1)(i)^{j-2} ,$$

$$u_{ij} = (j-1)(i)^{j-2}y_i + (i)^{j-1}y'_i .$$

The corresponding corrector matrix Δ_c can be obtained by replacing the first row in (6.2.24) by the last row, with ν replaced by $\nu+1$.

The equivalent corrector formula to (6.2.21) is given as

$$y_{n+2}^2 = 2y_{n+1}y_{n+2} + (y_{n+1}^2 - h^2 y'_{n+1}y'_{n+2}) . \tag{6.2.26}$$

Exercise. Show that the predictor formula for $\mu = 1$, $\nu = 2$ can be derived as

$$y_{n+1} = \frac{y_n^2(3y_{n+1}+hy'_{n+1}) + y_{n+1}^2(2hy'_n-3y_n)}{y_n(4y_n-5y_{n+1}+hy'_{n+1}) + y_{n+1}(y_{n+1}+2hy'_n) - 2h^2 y'_n y'_{n+1}} . \tag{6.2.27}$$

The predictor formula for the case $\mu = \nu-1$ can generally be obtained from the following constraints:

$$E_{\mu,\nu}(x_{n+j}) = 0 , j = 0(1)\nu \tag{6.2.28}$$

and

$$E'_{\mu,\nu}(x_{n+j}) = 0 , j = 0(1)\nu - 1 ,$$

while those of the corrector formula require the constraints

$$E_{\mu,\nu}(x_{n+j}) = 0 \ , \ j = 0(1)\nu \tag{6.2.29}$$

and

$$E'_{\mu,\nu}(x_{n+j}) = 0 \ , \ j = 1(1)\nu \ .$$

The corresponding corrector formula for (6.2.27) is given as

$$Cy_{n+1}^2 + Dy_{n+1} - E = 0 \ , \tag{6.2.30}$$

where

$$C = 3y_{n+1} - 4y_n - hy'_{n+1} \ ,$$

$$D = 5y_n y_{n+1} - 3y_{n+1}^2 + hy_n y'_{n+1} \ , \tag{6.2.31}$$

$$E = -y_n y_{n+1}^2 + 2h(y_n - y_{n+1})y_{n+1}y'_{n+2} + 2h^2 y_n y'_{n+1} y'_{n+2} \ .$$

6.3. The Inverse Polynomial Methods

Although the integration formulas based on rational approximations are quite effective in the neighborhood of singularities, the derivation of these formulas, and even the formulas themselves, are rather too involved and complicated. These formulas do not subject themselves to easy analysis and computer applications.

Fatunla (1982a) suggested the adoption of a rational function (6.2.11a) of a special kind whereby $\mu = 0$. Here, the theoretical solution $y(x)$ to the initial value problem is locally approximated by

$$F_k(x) = \frac{A}{1 + \displaystyle\sum_{j=1}^{k} a_j x^j} \ , \ k \ge 1 \ , \tag{6.3.1}$$

where A, a_j are real coefficients.

From (6.3.1) we now define the error equation $E_k(x)$ as

$$E_k(x) = (1 + \sum_{j=1}^{k} a_j x^j)y(x) - A \ , \tag{6.3.2}$$

which, on differentiating with respect to x, gives

$$E'_k(x) = (1 + \sum_{j=1}^{k} a_j x^j)y'(x) + (\sum_{j=1}^{k} ja_j x^{j-1})y(x) \ . \tag{6.3.3}$$

The imposition of the constraints

$$E_k(x_{n+i}) = 0 \ , \ i = 0(1)k \ ,$$

combined with the transformation (6.2.17), leads to the following integration formula:

$$y_{n+k} = \frac{y_n}{(1+\sum\limits_{j=1}^{k} k^j a_j)} \ . \tag{6.3.4}$$

It remains to obtain the numerical values of the components of the k-vector $a = (a_1, a_2, ..., a_k)^T$. This is achieved by ensuring that the interpolating function (6.3.1) satisfy the differential equation at k points $\{x_{n+j}, j = 0(1)k-1\}$. This implies

$$E'(x_{n+i}) = 0 \ , \ i = 0(1)k-1 \ . \tag{6.3.5}$$

Adopting the transformation (6.2.17) in (6.3.5) and replacing y_i' by hy_i' yields the following linear system of dimension k:

$$Ra = b \ , \tag{6.3.6}$$

where R is a k by k matrix with elements

$$R_{ij} = hi^j y_i' + ji^{j-1} y_i \ , \ i = 0(1)k-1; \ j = 1(1)k \ , \tag{6.3.7}$$

and b is a k-vector whose i-th element is

$$b_i = -hy_i' \ , \ i = 0(1)k-1 \ . \tag{6.3.8}$$

The System (6.3.6) has a unique solution if

$$\Delta = \det(R) \neq 0 \ . \tag{6.3.9}$$

In the event that $\det(R) = 0$, there is a strong indication of the existence of a singularity, and we can overstep this singularity by adjusting the stepsize. The singularity can be obtained from the poles of $F_k(x)$.

We now consider specific cases of the integration formula (6.3.4). Setting $k = 1$ in (6.3.4) gives the one-step integration formula

$$y_{n+1} = \frac{y_n}{(1+a_1)} \ , \tag{6.3.10}$$

with a_1 obtained from (6.3.6) as

$$R_{01} a_1 = b_0 \ . \tag{6.3.11}$$

From (6.3.7) we have

$$R_{01} = y_0 \ , \tag{6.3.12}$$

and from (6.3.8)

$$b_0 = -hy_0' \ . \tag{6.3.13}$$

Inserting (6.3.12) and (6.3.13) into (6.3.11) gives

$$a_1 = \frac{-hy_0'}{y_0} . \tag{6.3.14}$$

This, in (6.3.10), implies

$$y_{n+1} = \frac{y_n^2}{y_n - hy_n'} , \tag{6.3.15}$$

whose local truncation t_{n+1} error can be obtained as

$$t_{n+1} = \frac{(\frac{1}{2}y_n y_n' - y_n^2)}{(y_n - hy_n')} h^2 , \quad |y(x)| + |y'(x)| \neq 0 , \tag{6.3.16}$$

which suggests that (6.3.15) is at least of order $p \geq 1$, provided $|y_n| \neq 0$. In the event that y_n vanishes, the meshsize h should be adjusted. Systems of ODEs are treated component-wise.

Example. Solve the IVP in Example 6.1 using the nonlinear scheme (6.3.15).

Exercise. Derive the nonlinear integration formula (6.3.4) for the case $k = 2$. Then generate the numerical solution to Example 6.1 and compare results with those for $k = 1$.

Numerical solutions to Example 6.1 were generated with the nonlinear multistep formulas (6.3.4)—(6.3.8) for stepnumber $1 \leq k \leq 5$ whose details are given in Table 6.5.

Table 6.3. Numerical Solution to Problem (6.1) at $x = 0.75$
Method of Order One (6.3.15); Exact Solution: 28.2382528501

h	$y(x, h)$	$y(x, h) - y(x)$
0.05	57.2685650897	29.0303122396
0.025	37.5949476942	9.35669484408
0.0125	32.2111603071	3.9729074569
0.00625	30.0854849029	1.8472320528
0.003125	29.1306038013	0.8923509511

Table 6.4. Numerical Solution to Problem (6.1) at $x = 1.0$
Formula (6.3.15); Exact Solution: -4.58803782498

h	$y(x, h)$	$y(x, h) - y(x)$
0.05	-4.24590051301	0.34213731
0.025	-4.41346816233	0.174569663
0.0125	-4.49984309474	0.088194730
0.00625	-4.54370851518	0.044329310
0.003125	-4.56581461596	0.022223209

Table 6.5. Errors in Nonlinear Multistep Methods

x	Theoretical Solution	k = 1	k = 2	k = 3	k = 4	k = 5
0.0	1.000000000					
0.10	1.233048880	1.228(-2)	-8.277(-4)			
0.20	1.508497647	2.919(-2)	-1.894(-3)	2.298(-4)	-5.953(-5)	
0.30	1.895765123	5.580(-2)	-3.520(-3)	5.507(-4)	-5.434(-5)	2.303(-5)
0.40	2.464962757	1.045(-1)	-6.473(-3)	6.726(-4)	-1.685(-4)	2.408(-5)
0.50	3.408223442	2.134(-2)	-1.312(-2)	1.754(-3)	-2.238(-4)	7.234(-5)
0.60	5.331855223	5.562(-2)	-3.405(-2)	4.206(-3)	-7.541(-4)	1.272(-4)
0.65	7.340436375	1.106(0)	-6.664(-2)	8.379(-3)	-1.509(-3)	2.338(-4)
0.70	11.681373800	3.092(0)	-1.762(-1)	2.016(-2)	-3.322(-3)	7.849(-4)
0.75	28.238252850	2.903(1)	-1.074(0)	1.181(-1)	-1.893(-2)	4.902(-3)
0.80	-68.479668346	3.776(1)	-8.284(0)	7.037(-1)	-1.282(-1)	2.622(-2)
0.90	-8.687629546	1.166(0)	1.952(-1)	-6.093(-3)	-2.108(-2)	6.158(-3)
1.00	-4.588037825	3.421(-1)	2.340(-1)	9.291(-3)	1.610(-2)	1.054(-2)

Fatunla (1986) modified the automatic (polynomial/rational) extrapolation code DIFEX1 (Deuflhard 1983, 1985) to accommodate (6.3.15) as a basic integrator. This provides a variable step, variable order code DIFEX2. As is evident from Table 6.6, both the Gragg-Neville-Aitken polynomial and the Gragg-Bulirsch-Stoer rational extrapolation schemes coded in DIFEX1 are inefficient in the neighborhood of the singularity $x = \pi/4$ of the Example 6.1. This is simply because the basic integrator (Gragg's modified midpoint rule) with an error expansion in h^2 is based on polynomial interpolation, and, thus, requires more than the allowable maximum number of five stepsize reductions stipulated in the code.

However, the code DIFEX2, whose basic integrator is (6.3.15) and whose asymptotic error expansion is in h, can effectively and efficiently cope with singular IVPs, as is evident from Table 6.7. If a low degree of accuracy is desired, (6.3.15) combined with polynomial extrapolation is more efficient, but for a high degree of accuracy, (6.3.15) combined with rational extrapolation is recommended. This is evident from the results obtained for error tolerance $\varepsilon = 10^{-7}$ (i.e., $q = 7$) in Tables 6.6 and 6.7.

6.4. Local Error Estimates in Automatic Codes for Discontinuous Systems

Gear and Østerby (1984) suggested the use of local error estimators (in the absence of a switching function) to detect and locate a point of discontinuity. This was incorporated into an existing automatic code--GEAR (Hindmarsh 1974). The algorithm also provides estimates of the magnitude and order of the

Table 6.6.

Performance of Nonstiff Extrapolation Code DIFEX1 (Deuflhard 1983, 1985)
on the Singular IVP $y' = 1 + y^2$, $y(0) = 1$ $0 \le x \le 1$
$h_0 = 0.25$, $h_{max} = 1.0$, Maximum Extrapolation 6

$\varepsilon = 10^{-q}$ q	Modified Midpoint (+) Rational Extrapolation Gragg-Bulirsch-Stoer DIFEX1			Modified Midpoint (+) Polynomial Extrapolation Gragg-Neville-Aitken DIFEX1		
	Termination Point x	No. of Steps	No. of Fn. Eval.	Termination Point x	No. of Steps	No. of Fn. Eval.
1	0.904585413	17	299	0.785492433	132	1710
2	0.786585246	88	880	0.786472360	88	880
3	0.785412676	97	1843	0.785399251	103	1975
4	0.785399984	203	5057	0.785398887	208	5182
5	0.785398693	166	3190	0.785399861	164	3140
6	0.785398201	179	2843	0.785398171	153	5045
7	0.785398166	289	4471	0.785398170	180	5936

Table 6.7

Performance of New Extrapolation Codes DIFEX2 (Fatunla 1986c)

on Singular IVP $y' = 1 + y^2$, $y(0) = 1$, $0 \leq x \leq 1$.

$h_0 = 0.25$, $h_{max} = 1.0$, Maximum Extrapolation 6

Theoretical Solution: $y(x) = \tan(x + \pi/4)$

Exact Solution: -4.588037825

	Inverse Euler (+) Rational Extrapolation			Inverse Euler (+) Polynomial Extrapolation		
$\varepsilon = 10^{-q}$ q	y	No. of Steps	No. of Fn. Eval.	y	No. of Steps	No. of Fn. Eval.
1	-4.69169062	5	41	-4.66101122	5	41
2	-4.58611369	5	65	-4.58357207	3	47
3	-4.58834175	4	72	-4.58788415	3	55
4	-4.58804399	4	108	-4.58800517	3	63
5	-4.58802489	4	126	-4.58803957	3	105
6	-4.58803827	4	162	-4.58803846	3	105
7	-4.58803780	4	188	-4.58803822	691	5107

discontinuity. Gear (1980d) provides an efficient vehicle to restart integration beyond the point of discontinuity.

The four computational steps adopted are

(a) detect the presence of discontinuity,

(b) locate the point of discontinuity,

(c) pass the point of discontinuity, and

(d) restart the integration process.

The presence of a discontinuity is identified if where TOL is the allowable error tolerance, p is the error order of the method, and le_n is an estimate of the local truncation error.

$$ |\frac{\text{TOL}}{le_n}|^{\frac{1}{(p+1)}} , \tag{6.4.1} $$

The discontinuity is normally located to a desired accuracy by repeatedly halving the stepsize.

As a result of the discontinuity, the back values of y and f may no longer be valid, and, hence, cannot be used in a linear multistep formula. The explicit Euler method can be used to predict, while the implicit Euler or the trapezoidal rule is used as the corrector formula. Alternatively, the Runge-Kutta starter proposed in Ellison (1981) could be adopted.

7. EXTRAPOLATION PROCESSES AND SINGULARITIES

7.1. Introduction

The concept of extrapolation has played a significant role in numerical mathematics. It can be exploited to obtain a numerical estimate for the physical constant π by observing that the perimeter of a regular n-sided polygon inscribed in a circle with unit diameter is given as

$$s(n) = n \sin \frac{\pi}{n} , \qquad (7.1.1)$$

and that this can be made to be as close as possible to the circumference of the circle by choosing n sufficiently large; i.e., $\lim\limits_{n \to \infty} s(n) = \pi.$

With a variable parameter h satisfying

$$nh = 1 , \qquad (7.1.2)$$

we can recast (7.1.1) as

$$y(h) = \frac{1}{h} \sin \pi h .$$

Adopting the MacLaurin's series expansion for $\sin \pi h$ in the last equation leads to the following asymptotic error expansion in h:

$$y(h) = \pi + \sum_{r=1}^{\infty} a_r h^{2r} , \qquad (7.1.3)$$

with the coefficients a_r given by

$$a_r = (-1)^r \frac{\pi^{2r+1}}{(2r+1)!} . \qquad (7.1.4)$$

Equation (7.1.3) satisfies

$$y(0) = \pi .$$

From Equations (7.1.3) and (7.1.4), we derive

$$y(h, \frac{h}{2}) = \frac{2^2 y(\frac{h}{2}) - y(h)}{2^2 - 1}$$

$$(7.1.5)$$

$$= \pi - \frac{\pi^5}{480} h^4 + O(h^6) ,$$

which gives a better estimate for π than either $y(h)$ or $y(\frac{h}{2})$. Further improved accuracy can be obtained by the following linear combination of $y(h)$:

$$y(h, \frac{h}{2}, \frac{h}{4}) = \frac{4^2 y(\frac{h}{2}, \frac{h}{4}) - y(h, \frac{h}{2})}{4^2 - 1}$$

$$(7.1.6)$$

$$= \pi + O(h^6) .$$

The points

$$\{(h, y(h)), (\frac{h}{2}, y(\frac{h}{2})), (\frac{h}{4}, y(\frac{h}{4})), (\frac{h}{8}, y(\frac{h}{8}))\}$$

are in the solution space, and they constitute elements of the *zero*-th column of the extrapolation table.

The elements of column j, $j \geq 1$ are obtained by polynomial interpolations of order $(j-1)$ of points in the solution space by successive linear interpolation of elements in the proceeding $j-1$ columns.

Exercise. 1. Obtain $y(h, h/2, ..., h/2^8)$ with $h = 1/6$. 2. Develop an alternative scheme to estimate π by circumscribing a regular polygon on a circle with unit diameter.

In general, a physical quantity whose exact value is $y(x)$ can be approximated using a numerical algorithm that involves a parameter $h > 0$ in the following way:

Extrapolation Table 7.1

$$y(h)$$

$$y(\tfrac{h}{2}) \quad y(h, \tfrac{h}{2})$$

$$y(\tfrac{h}{4}) \quad y(\tfrac{h}{2}, \tfrac{h}{4}) \quad y(h, \tfrac{h}{2}, \tfrac{h}{4}) \qquad\qquad (7.1.7)$$

$$y(\tfrac{h}{8}) \quad y(\tfrac{h}{4}, \tfrac{h}{8}) \quad y(\tfrac{h}{2}, \tfrac{h}{4,}\tfrac{h}{8}) \quad y(h, \tfrac{h}{2}, \tfrac{h}{4}, \tfrac{h}{8})$$

$$y(x, h) = y(x) + \sum_{r=1}^{\infty} a_r(x) h^{\alpha_r} , \qquad\qquad (7.1.8)$$

$$\alpha_r = \gamma r \qquad\qquad (7.1.9)$$

for some positive integer γ ($\gamma = 2$ in (7.1.3)).

A sequence $\{y(x, h_r) \,|\, r = 0, 1, 2,...\}$ is generated with a strictly decreasing sequence of parameters h_r

$$\{h_r \,|\, h_0 = h, \ h_r > h_{r+1} , \ r = 0, 1, 2,...\} . \qquad\qquad (7.1.10)$$

It is obvious that $y(x, 0) = y(x)$, but the trouble is that, in the numerical approximation of $y(x)$, it is impossible to adopt $h = 0$. Instead, we can extrapolate by a polynomial or rational function to the point $(0, y(x, 0))$ from the points

$$\{(h_r, y(x, h_r)) \,|\, r = 0(1)M\}$$

in the solution space. The quantities $\{y(x, h_r) \,|\, r = 0(1)M\}$ constitute the zero-th column of the extrapolation Table 7.1.

7.2. Generation of the Zero-th Column of Extrapolation Table

Our main concern is to make use of this versatile and important tool called extrapolation in order to increase the order of the numerical results generated by the lower order discretization schemes for the IVP (2.1.4) without seriously jeopardizing the stability properties of the original method.

The lower order schemes adopted for generating the zero-th column of the extrapolation table with elements $T_{00}, T_{10}, T_{20}, ..., T_{M0}$ include

(a)   the Euler's method (3.2.2),

(b)   the generalized trapezoidal scheme (3.4.2),

(c)   the modified midpoint rule (7.2.11),

(d)   the inverse Euler scheme (3.2.9), or the implicit Euler scheme (3.2.7),

(e)   implicit midpoint rule (3.4.16), and

(f)   the semi-implicit midpoint rule specified in (7.2.19), proposed by Bader and Deuflhard (1983).

The choice of method will depend, to a large extent, on the following factors:

(i)   The nature of the problem. For non-stiff initial value problems, the Euler's scheme and the modified midpoint rule are quite suitable, with preference to the midpoint rule, as it possesses an asymptotic error expansion in h^2. The generalized trapezoidal rule is suitable for stiff systems, as its A-stability property is retained provided global extrapolation (i.e., extrapolation tables are generated simultaneously at all the meshpoints in the integration interval) is employed and all the elements of the zero-th column are generated accurately; i.e., iteration to convergence is adopted. Then there is an error expansion in h^2.

Being A-stable and having an error expansion in h^2, the implicit midpoint rule is particularly well suited to stiff initial value problems. It does not require global extrapolation (Lindberg 1973). The inverse Euler's rule (6.3.15) is particularly well suited to IVPs containing singularities. The snag is that the scheme is only component-applicable to systems (cf. Lambert 1973a, p. 218).

(ii)   There is a need for the fast convergence of the extrapolation process produced by an asymptotic error expansion (7.1.8) in h^2; i.e., $\gamma_r = 2r$, as in the case of the trapezoidal scheme, the modified midpoint rule, and the semi-implicit midpoint rule.

Let h be the integration meshsize and obtain the following sequence of meshsizes as a special case of (7.1.10):

$$\{\, h_r \mid h_r = \frac{h}{2^{r+1}} \,,\, r = 0(1)M \,\} \,, \tag{7.2.1}$$

where M is the maximum allowable order of extrapolation.

From (7.2.1) we generate the following sequence of integers:

$$\{N_r \mid N_r = \frac{h}{h_r} \,,\, r = 0(1)M \} \,. \tag{7.2.2}$$

With (x_n, y_n) available, $\{t_s\}$ is a sequence of meshpoints given as

$$\{t_s \mid t_s = x_n + s h_r \,;\, s = 0(1)N_r \} \,. \tag{7.2.3}$$

It should be noted that, from (7.2.3),

$$t_0 = x_n \,,\, t_{N_r} = x_{n+1} \,.$$

Euler's Scheme

With $z_0 = y_n$, an m-tuple representing the numerical solution at $x = x_n$, the r-th element T_{r0} of the zero-th column of the extrapolation table is generated as follows:

$$z_{s+1} = z_s + h_r f(t_s, z_s), \ s = 0(1)N_r - 1 \tag{7.2.4}$$

and

$$T_{r0} = z_{N_r} = y(x_{n+1}, h_r) . \tag{7.2.5}$$

All the $M+1$ elements of the zero-th column are generated by adopting each of the meshsizes specified in (7.2.1). The numerical approximations (7.2.5) have an asymptotic error expansion in h; i.e., $\alpha_r = r$ in (7.1.8).

The Generalized Trapezoidal Rule

It was observed earlier that the trapezoidal rule (3.4.2) is more accurate and possesses better stability properties than Euler's scheme (3.2.2), (cf. Dahlquist 1963). Besides, the trapezoidal scheme possesses an error expansion in even powers of h, provided it is implemented in the iteration to convergence mode (cf. Gragg 1964). It is thus a more attractive candidate for extrapolation.

Richardson (1927) and Romberg (1955) combined the trapezoidal scheme with extrapolation for the evaluation of the definite integral

$$y(x) = \int_a^b f(x)dx , \tag{7.2.6}$$

whose discretization yields

$$y(x, h) = h[\frac{1}{2}f(a) + f(a+h) + ... + f(b-h) + \frac{1}{2}f(b)] , \tag{7.2.7}$$

which does have an error expansion in h^2.

The generalized trapezoidal rule for the numerical solution of the initial value problem (2.1.4) is given as

$$y_{n+1} = y_n + \frac{h}{2}[f(x_n, y_n) + f(x_{n+1}, y_{n+1})] , \tag{7.2.8}$$

which has to be solved accurately in order to possess property (7.1.8) with $\alpha_r = 2r$. A Newton scheme is called into play to accomplish this.

In order to preserve the A-stability property of the trapezoidal rule, it is imperative that global extrapolation, rather than local extrapolation, be adopted. In case of local extrapolation, only one extrapolation table is generated to obtain numerical estimate y_{n+1} to $y(x_{n+1})$. As for global extrapolation, N extrapolation tables are generated simultaneously; one at each of the meshpoints x_n, $n = 1(1)N$, $H = (b-a)/N$.

With a uniform meshsize h given, the meshpoints

$$\{x_n \,|\, x_n = a + nh \, , \, n = 0(1)M\} \tag{7.2.9}$$

are obtained where $M = (b-a)/h$. An extrapolation table is required for each of the meshpoints x_n, $n = 1(1)M$. The zero-th column of these M tables can be generated as follows: Using equations (7.2.1)—(7.2.3), generate

$$z_0 = y_n \, ,$$

$$z_{s+1} = z_s + \frac{h_r}{2} [f\,(t_s, z_s) + f\,(t_{s+1}, z_{s+1})] \, , \ \ s = 0(1)N_r{-}1 \, , \tag{7.2.10}$$

$$T_{r0}^{(n)} = z_{N_r}$$

for $r = 0(1)M$, and $n = 1(1)N$.

The extrapolation process can now be implemented at each of the N meshpoints x_n, $n = 1(1)N$. Without computational experience, the implementation process may demand a sizeable amount of core storage.

Modified Midpoint Rule (Gragg 1964, 1965)

Although the trapezoidal rule has an h^2 error expansion, its main snags are its implicitness and the requirement that this be solved exactly at each meshpoint.

Gragg (1965) proposed an alternative scheme with stepnumber two, which possesses the same desirable property of an asymptotic error expansion in h^2.

Consider the interval $[x_n, x_{n+1}]$, and, using equations (7.2.1)—(7.2.3), generate the elements of the zero-th column $T_{r0}, r = 0(1)M$ as follows:

$$z_1 = z_0 + h_r f(t_0, z_0) \, ,$$

$$z_{s+1} = z_{s-1} + 2h_r f\,(t_s, z_s) \, , \, s = 1(1)N_r{-}1 \, , \tag{7.2.11}$$

$$T_{r0} = \frac{1}{2}[z_{N_r-1} + z_{N_r} + hf\,(t_{N_r}, z_{N_r})] = y\,(x_{n+1}, h_r) \, , \, r = 0(1)M \, .$$

The last equation in (7.2.11) is a smoothing procedure designed to eliminate the instability of the midpoint rule. Each approximate solution $y(x_{n+1}, h_r)$ retains the asymptotic error expansion in h^2 of the the the midpoint rule. Shampine and Baca (1983) showed that there is no need to remove the weakly-stable component by smoothing; extrapolation will do it automatically. A cheaper smoothing scheme is also proposed, and smoothing is also exploited to increase the robustness of extrapolation codes (cf. Shampine 1983b).

Gragg (1965) further established that, for any k-step linear multistep method (5.1.1) to possess an error expansion in h^2, it has to be symmetric in the sense that its first and second characteristic polynomials satisfy the relation

$$\rho(r) + r^k \rho(\frac{1}{r}) = \sigma(r) - r^k \sigma(\frac{1}{r}) \, . \qquad (7.2.12)$$

The midpoint rule is the simplest of such methods.

Inverse Euler Scheme (3.2.9)

The formulation of Euler's scheme, the trapezoidal scheme, and the modified midpoint rule is essentially that of polynomial interpolation. Consequently, such schemes perform poorly in the neighborhood of a singularity of any kind. Better results can be obtained in the neighborhood of a singularity if the zero-th column of the extrapolation table is generated with the inverse Euler scheme (cf. Fatunla 1982a)

$$y_{n+1} = \frac{y_n^2}{y_n - h y_n'} \, , \qquad (7.2.13)$$

developed in Section 6.3. With h_r, N_r, and t_s given by (7.2.1), (7.2.2), and (7.2.3), respectively, the zero-th column of the extrapolation table can be generated as follows:

$$z_0^i = y_n^i \, , \, i = 1(1)M \, ,$$

$$z_{s+1}^i = \frac{(z_s^i)^2}{z_s - h_r(z_s^i)'} \, , \, s = 0(1)N_r - 1 \, , \, i = 1(1)m \, , \qquad (7.2.14)$$

$$T_{r_0}^i = z_{N_r}^i = y^i(x_{n+1}, h_r) \, ,$$

where the superscript i denotes the i-th component of vectors y_n, z_s. As mentioned earlier, this scheme is component-applicable to systems of ordinary differential equations. However, component-wise application may destroy stability.

Implicit Midpoint Rule

In the program package IMPEX, Lindberg (1973) adopted the implicit midpoint rule as the basic integrator. It was specified as

$$y_{n+1} = y_n + hf(x_{n+1/2}, \frac{y_n + y_{n+1}}{2}) \, .$$

It is implicit, has asymptotic error expansion in h^2, and retains its A-stability properties under global extrapolation, which places it at an advantage over the trapezoidal scheme.

Using Equations (7.2.1)—(7.2.3), the zero-th column of the extrapolation table is generated as follows:

$$z_0 = y_n \, ,$$

$$z_{s+1} = z_s + h_r f\left(x_{n+1/2}, \ \frac{z_s + z_{s+1}}{2}\right),$$ (7.2.15)

$$T_{r0} = z_{N_r} = y(x_{n+1}, h_r) \ ; \ s = 0(1)N_r - 1 \ , \ r = 0(1)M \ .$$

Semi-Implicit Midpoint Rule

Bader and Deuflhard (1983) proposed the following semi-implicit scheme for stiff initial value problems:

$$y' = f(y) .$$ (7.2.16)

With

$$J = f_y(y)$$ (7.2.17)

as the Jacobian of $f(y)$, let

$$\bar{f}(y) = f(y) - Jy .$$ (7.2.18)

With a specified integer $2r$, each subintegration from x_n to $x_n + H$ (meshsize $h_r = H/2r$) with the semi-implicit midpoint is affected as follows:

$$\eta_0 = y_n ,$$

$$\eta_1 = (I - h_r J)^{-1}[y_n + h_r \bar{f}(y_n)] ,$$

$$\eta_{k+1} = (I - h_r J)^{-1}[(I + h_r J)\eta_{k-1} + 2h_r \bar{f}(\eta_k)] , \ k = 1(1)2r ,$$ (7.2.19)

$$s_{2r} = \frac{1}{2}(\eta_{2r-1} + \eta_{2r+1}) ,$$

$$= T_{r0} = y(x_n + H, h_r) ,$$

the last equation being a smoothing procedure. Higher order approximations of $y(x_n + H)$ are obtained by performing polynomial extrapolation (7.3.2) in h^2; i.e., $\gamma = 2$.

If $J = 0$ in (7.2.19), the resultant algorithm is identical to the modified midpoint rule given by (7.2.11), suitable for nonstiff initial value problems. The only variation is the smoothing procedure, which apparently has improved stability characteristics in that it provides stronger damping at infinity. However, the semi-implicit midpoint rule (7.2.19) demands the following additional computational efforts over the Gragg midpoint rule (7.2.11):

(a) The computation of the Jacobian J for each step H. If the whole problem is clearly nonstiff, the explicit rule is appropriate; i.e., $J = 0$.

(b) r Gaussian decompositions of $I - h_r J$ for each step $x_n \to x_n + H$.

(c) The solution of r sets of $2r + 1$ linear systems of algebraic equations.

7.3. Polynomial and Rational Extrapolation

In Section 3.2, the Richardson's extrapolation to the limit was applied to both the Euler and the inverse Euler schemes. It was also adopted in Section 4.1 to obtain the local truncation errors for the Runge-Kutta scheme.

Having generated the $M+1$ points $\{(h_r, y(x_{n+1}, h_r)) \mid r = 0(1)M\}$ in the solution space, it is possible to obtain a polynomial $P_M(h)$ of degree M, or a rational function $R_{\mu,\nu}(h)$ (μ denotes the degree of numerator, and ν denotes the degree of the denominator), which satisfies

$$P_M(h_r) = R_{\mu,\nu}(h_r) = T_{r,0} , r = 0(1)M . \tag{7.3.1}$$

Polynomial Extrapolation

We have shown the existence of such a unique polynomial $P_M(h)$ in Section 1.2. Aitken (1932) and Neville (1934) each independently proposed a scheme whereby the polynomial $P_M(h)$ is generated recursively by linear interpolation. The entries of the remaining M columns of the extrapolation Table 7.2 are derived column by column, as follows:

$$T_{r,0} = y(x_{n+1}, h_r) , r = 0(1)M ,$$

$$\tag{7.3.2}$$

$$T_{r,s} = T_{r+1,s-1} + \frac{T_{r+1,s-1} - T_{r,s-1}}{(\dfrac{h_r}{h_{r+s}})^{\gamma} - 1} , s = 1(1)M , r = 0(1)M-s .$$

$$\gamma = \begin{cases} 1 , & \text{if } T_{r0} \text{ has asymptotic error expansion in } h \\ 2 , & \text{if } T_{r0} \text{ has asymptotic error expansion in } h^2 \end{cases}$$

Table 7.2. Extrapolation Table for Fixed Order M

T_{00}					
T_{10}	T_{01}				
T_{20}	T_{11}	T_{02}			
$\vdots$	$\vdots$	$\vdots$			
$T_{M-1,0}$	$T_{M-2,1}$			$T_{0,M-1}$	
T_{M0}	$T_{M-1,1}$	$T_{M-2,2}$	$\cdots$	$T_{1,M-1}$	$T_{0,M}$

Example. Solve the initial value problem $y' = 1 + y^2$, $y(0) = 1$ over the interval $0 \le x \le 0.25$, using $h = 0.25$ and order of extrapolation $M = 4$. Adopt the inverse Euler polynomial extrapolation with $\gamma = 1$.

The error estimate E_M for the computed result in the step $x \to x + H$ is given by Bulirsch and Stoer (1966b) as

$$E_M = T_{0,M} - y(x)$$

$$\le h_0 D_M (h_0 \, h_1 \dots h_M)^\gamma , \tag{7.3.3}$$

which indicates that the error decreases with increasing γ.

The equation

$$h_0^* = h_0 \left(\frac{h_0 * \text{EPS}}{E_M} \right)^{1/2M} \tag{7.3.4}$$

can be adopted as a stepsize adjustment if it happens that there were convergence difficulties.

Stetter (1973) proved that polynomial extrapolation possesses weak stability characteristics compared to other methods. Shampine and Baca (1984) established that the result of polynomial extrapolation of the explicit midpoint rule corresponds to forming a highly-structured family of explicit Runge-Kutta formulas, so a code based on it represents a variable order R-K code.

Table 7.3. Solutions at $x = 0.25$
Exact Solution: $y(x) = \tan(x + \pi/4)$

1.8028169014			
1.7378729128	1.6729289242		
1.7104996963	1.6831264798	1.6865256650	
1.6978418056	1.6851839150	1.6858697276	1.6857760212

Inverse Euler (7.2.14) + Polynomial Extrapolation (7.3.2)

Error Table 7.4

$h_0 D_0 h_0^\gamma$			
$h_0 D_0 h_1^\gamma$	$h_0 D_1 (h_0 h_1)^\gamma$		
$h_0 D_0 h_2^\gamma$	$h_0 D_1 (h_1 h_2)^\gamma$	$h_0 D_2 (h_0 h_1 h_2)^\gamma$	
$h_0 D_0 h_3^\gamma$	$h_0 D_1 (h_2 h_3)^\gamma$	$h_0 D_2 (h_1 h_2 h_3)^\gamma$	$h_0 D_3 (h_0 h_1 h_2 h_3)^\gamma$

Rational Extrapolation

Experience has shown that extrapolation based on rational functions is superior to polynomial extrapolation particularly in the neighborhood of a singularity. Bulirsch and Stoer (1966a) adopted the rational function $R_{\mu,\nu}(h)$ that passes through the $M+1$ points $\{(h_r, y(x_{n+1}, h_r))\,|\,r = 0(1)M\}$ and is specified by

$$R_{\mu,\nu}(h) = \frac{(\sum\limits_{r=0}^{\mu}\alpha_r h^{\gamma_r})}{(\sum\limits_{r=0}^{\nu}\beta_r h^{\gamma_r})}\ , \tag{7.3.5}$$

with

$$\mu = [\frac{M}{2}]\ ,\ \nu = M - \mu\ . \tag{7.3.6}$$

The interpolating function thus satisfies the requirement

$$R_{\mu,\nu}(h_r) = T_{r0}\ ,\ r = 0(1)M\ ,$$

with h_r specified in (7.1.10). Bulirsch and Stoer asserted that the adoption of

$$H = \{h,\ \frac{h}{2},\ \frac{h}{3},\ \frac{h}{4},\ \frac{h}{6},\ \frac{h}{8},...\} \tag{7.3.7}$$

yields equal accuracy and requires less computational effort.

The subsequent columns of the extrapolation table are generated as follows:

$$T_{r,-1} = 0\ ,$$

$$T_{r,0} = y(x_{n+1},\ h_r)\ ,\ r = 0(1)M\ , \tag{7.3.8}$$

$$T_{r,s} = T_{r+1,s-1} + \frac{T_{r+1,s-1} - T_{r,s-1}}{(\dfrac{h_r}{h_{r+s}})^\gamma\,[1 - \dfrac{T_{r+1,s-1} - T_{r,s-1}}{T_{r+1,s-1} - T_{r+1,s-2}}] - 1}$$

for $s = 1(1)M,\ r = 0(1)M - s$.

Stoer (1961) established that $T_{r,s}$ is equivalent to the interpolating function

$$T_{r,s}(h) = \begin{cases} (\sum\limits_{j=0}^{s/2} a_j h^{2j})/(\sum\limits_{j=0}^{s/2} b_j h^{2j})\ , & \text{seven}\ , \\[6pt] (\sum\limits_{j=0}^{[s/2]} a_j h^{2j})/(\sum\limits_{j=0}^{[s/2]+1} b_j h^{2j})\ , & s\,\text{odd}\ . \end{cases} \tag{7.3.9}$$

Wuytack (1971, 1974) constructed $T_{r,s}$ in a more efficient manner—in terms of continuous fractions. He computed only alternate terms (i.e., $r+s$ even), and his approach demands fewer arithmetic operations.

Gragg (1965) showed that asymptotic expansions in h^2 exist, provided that the method in question is symmetric; i.e.,

$$\rho(z) + z^k \rho(z^{-1}) = \sigma(z) - z^k \sigma(z^{-1}) = 0 , \qquad (7.3.10)$$

with the assumption that the starting values are exact. The use of implicit formulas is discouraged, as error expansions are valid only if exact solutions are obtained at each integration step.

Problem. Solve the last example using

(a) inverse Euler (7.2.14) + rational extrapolation (7.3.8) with $\gamma = 1$,

(b) Gragg-Bulirsch-Stoer, i.e., modified midpoint rule (7.2.11) + rational extrapolation (7.3.8) with $\gamma = 2$.

As anticipated, the Gragg-Bulirsch-Stoer extrapolation scheme is more accurate at $x = 0.25$, since the asymptotic error expansion is in h^2. The inverse Euler combined with rational extrapolation is slightly superior to the inverse Euler polynomial extrapolation and the Euler's scheme rational extrapolation.

Table 7.5a. $y' = 1 + y^2$, $y(0) = 1$, $0 \leq x \leq 0.25$, $h = 0.25$
Exact Solution $y(0.25) = 1.685796417$
Extrapolation Table; Inverse Euler Rational Extrapolation $\gamma = 1$

1.8028169014			
1.7378739128	1.6774452643		
1.7104996963	1.6839754186	1.6858494038	
1.6978418056	1.6853698785	1.6858026769	1.6857961912
			(7.2598×10^{-6})

Table 7.5b. Gragg-Bulirsch-Stoer $\gamma = 2$

1.6406250000			
1.6731318956	1.6842556897		
1.6825207564	1.6856738366	1.6857813466	
1.6849701171	1.6857881562	1.6857960729	1.6857963393
			(3.4425×10^{-7})

Table 7.5c. Euler's Scheme + Rational Extrapolation $\gamma = 1$

1.5703125000			
1.6195625218	1.6720018467		
1.6499448310	1.6814888520	1.6854629545	
1.6670804665	1.6845757643	1.6857435231	1.6857868712
			(5.2894×10^{-5})

The superiority of the inverse Euler combined with either the polynomial or rational extrapolation can be better appreciated in the neighborhood of a singularity. For the problem at hand, which has a singularity at $x = \pi/4$, we now attempt to adopt the four procedures to integrate this problem numerically in the interval $0 \le x \le 1$, with a uniform meshsize $h = 0.25$. The results of the numerical experiment are illustrated in Table 7.6.

Both

(a)   the Euler + rational extrapolation and

(b)   the Gragg-Bulirsch-Stoer extrapolation

perform poorly at $x \ge 0.75$ because the zero-th columns of the extrapolation tables are generated by the Euler and modified midpoint rules whose formulations are based essentially on polynomial interpolation.

The Gragg-Bulirsch-Stoer algorithm can be more efficiently and reliably implemented if the following recursive approach is adopted:

Firstly, Equation (7.3.8) is rearranged to give

$$T_{r,s} = T_{r,s-1} + \cfrac{(\dfrac{h_r}{h_{r+s}})^\gamma [1 - \dfrac{T_{r+1,s-1} - T_{r,s-1}}{T_{r+1,s-1} - T_{r+1,s-2}}](T_{r+1,s-1} - T_{r,s-1})}{(\dfrac{h_r}{h_{r+s}})^\gamma [1 - \dfrac{T_{r+1,s-1} - T_{r,s-1}}{T_{r+1,s-1} - T_{r+1,s-2}}] - 1} \; ;$$

i.e.,

Table 7.6. $y\prime = 1 + y^2$, $y(0) = 1$, $y(x) = \tan(x + \pi/4)$; $h = 0.25$, $M = 4$

x	0.25	0.50	0.75	1.00
Theoretical	1.685796417	3.408223442	28.238252850	-4.588037825
Inverse Euler Polynomial Extrap.	1.685802677	3.408247425	28.23948693	-4.587948676
Inverse Euler Rational Extrap.	1.685869727	3.408547782	28.264477616	-4.587322750
Euler Rational Extrap.	1.685743523	3.406978817	25.991812595	overflow
Gragg-Bulirsch-Stoer	1.685796073	3.408196987	27.315540959	overflow

$$T_{r,s} = T_{r,s-1} + \frac{(\frac{h_r}{h_{r+s}})^\gamma [T_{r,s-1}-T_{r+1,s-2}](T_{r+1,s-1}-T_{r,s-1})}{(\frac{h_r}{h_{r+s}})^\gamma [T_{r,s-1}-T_{r+1,s-2}] - (T_{r+1,s-1}-T_{r+1,s-2})}, \qquad (7.3.11)$$

on inserting

$$C_{r,s} = T_{r,s} - T_{r+1,s-1},$$

$$(7.3.12)$$

$$D_{r,s} = T_{r,s} - T_{r,s-1},$$

and

$$E_{r,s} = T_{r,s} - T_{r-1s}.$$

In (7.3.8) we obtain the following expression:

$$C_{r,s} = \frac{E_{r+1,s-1} D_{r+1,s-1}}{(\frac{h_r}{h_{r+s}})^\gamma C_{r,s-1} - D_{r+1,s-1}}. \qquad (7.3.13)$$

Equation (7.3.11) also reduces to

$$D_{r,s} = \frac{(\frac{h_r}{h_{r+s}})^\gamma C_{r,s-1} E_{r+1,s-1}}{(\frac{h_r}{h_{r+s}})^\gamma C_{r,s-1} - D_{r+1,s-1}}. \qquad (7.3.14)$$

The Gragg-Bulirsch-Stoer given by (7.3.8) then gives

$$C_{r0} = D_{r0} = y(x_{n+1}, h_r),$$

$$(7.3.15)$$

$$E_{r0} = y(x_{n+1}, h_r) - y(x_{n+1}, h_{r-1}).$$

Compute $C_{r,s}$, $D_{r,s}$, and $E_{r,s}$ from (7.3.13), (7.3.14), and (7.3.12), respectively, for $s = 1(1)M, r = 0(1)M-s$.

As

$$T_{r,s} = \sum_{r=1}^{r} D_{r,s-r}, \qquad (7.3.16)$$

it suffices only to save the elements of a column vector $C_{r,s-1}$ for $r = s(-1)0$ in computer memory.

Table 7.7

$$(0,0)$$
$$(1,0) \quad (0,1)$$
$$(2,0) \quad (1,1) \quad (0,2) \tag{7.3.17}$$
$$(3,0) \quad (2,1) \quad (1,2) \quad (0,3)$$
$$(4,0) \quad (3,1) \quad (2,2) \quad (1,3) \quad (0,4)$$

Stoer (1974) suggested a procedure for variable order variable step extrapolation schemes, but this has since been superseded by a much better scheme.

Deuflhard (1983) suggested a new technique where order and stepsize may vary simultaneously along the trajectory of each individual problem. This is based on the sub-diagonal error criterion of the form

$$e_{1,r} \approx T_{1,r} - T_{0,r+1} \le \varepsilon, \tag{7.3.18}$$

where ε is the prescribed local error tolerance, and $E_{1,r}$ is an estimate of the local error. Under sufficient conditions, the higher order approximation is more accurate than the lower order approximation; i.e.,

$$e_{1,r+1} \prec e_{1,r}. \tag{7.3.19}$$

If (7.3.18) holds, local extrapolation is adopted by setting

$$y(x_{n+1}, h) = T_{0,r+1}. \tag{7.3.20}$$

The stepsize for the next integration step is taken as

$$h_{new} = \left[\frac{\varepsilon}{e_{1,r}} \right]^{1/p_r} * h, \tag{7.3.21}$$

and the order of accuracy of solution is

$$p_r = \begin{cases} r, & \text{if error expansion is in } h \\ 2r, & \text{if error expansion is in } h^2. \end{cases} \tag{7.3.22}$$

If, however, on the completion of the extrapolation table, (7.3.18) is violated for $r = M-1$, then the stepsize is halved:

$$h_{new} = 0.5 h_{old}. \tag{7.3.23}$$

7.4. Convergence and Stability Properties of Extrapolation Processes

Each diagonal element $T_{0,s}$ of the extrapolation Table 7.7 can be expressed as a linear combination of the elements of the zero-th column; i.e.,

$$T_{0,s} = \sum_{r=0}^{s} \alpha_{s,s-r} T_{r0} \tag{7.4.1}$$

for some finite constants $\alpha_{s,s-r}$.

The following two theorems due to Laurent (1963) and Gragg (1965), guarantee the convergence of the extrapolation process (Table 7.7):

Theorem 7.1 (Laurent 1963). *A necessary and sufficient condition that*

$$\lim_{M \to \infty} T_{0,M} = y(x_{n+1}, 0) \tag{7.4.2}$$

for all $y(x_{n+1}, h)$ *continuous from the right at* $h = 0$ *is that*

$$\beta = \sup_{r \geq 0} \frac{h_{r+1}}{h_r} < 1. \tag{7.4.3}$$

In particular, Condition (7.4.3) implies the Toeplitz condition that the coefficients of (7.4.1) fulfill

$$C = \sup_{r \geq 0} \sum_{r=0}^{s} |\alpha_{s,s-r}| < \infty, \tag{7.4.4}$$

where constant C is a measure of the numerical stability of the extrapolation scheme.

Deuflhard (1983) has found it advantageous (in terms of efficiency) to use the slowest possible rate of increase, namely

$$\{N_r\} \equiv \{2, 4, 6, 8, 10, 12, 14, 16, 18, 20\},$$

rather than (7.2.2).

Theorem 7.2 (Gragg 1964, 1965). *Let* $y(x_{n+1}, h)$ *possess the asymptotic error expansion (7.1.8) such that*

$$\sup_{r \geq 0} \frac{h_{r+1}}{h_r} \leq \theta < 1. \tag{7.4.5}$$

Then as $r \to \infty$,

$$T_{r,s} - y(x_{n+1}, 0) = (-1)^s e_{s+1}(h_r \, h_{r+1} \, \cdots \, h_{r+s})^\gamma$$

$$+ O(h_r \, \cdots \, h_{r+s}) \gamma + 1. \tag{7.4.6}$$

In addition, if

$$0 < \delta \leq \inf_{r \leq 0} \frac{h_{r+1}}{h_r}, \tag{7.4.7}$$

then there exists constants E_s *such that for each* $s \geq 0$,

$$|T_{0r} - y(x_{n+1}, 0)| \leq E_{s+1}(h_0 \, h_1 \, \cdots \, h_r)^\gamma. \tag{7.4.8}$$

In the event that $e_s \neq 0$, $s \geq 1$, Equation (7.4.6) confirms that each column $s \geq 1$ of the extrapolation Table 7.7 converges faster than the preceding column $s-1$. Equation (7.4.8) asserts that the diagonal elements $T_{0,s}$ of Table 7.7

converge to $y(x_{n+1}, 0)$ faster than any of the columns of the extrapolation Table 7.7. In fact, under a mild restriction on the rate of growth of order constants in (7.1.8), the diagonal elements converge superlinearly to $y(x_{n+1}, 0)$, as

$$|T_{0,s} - y(x_{n+1}, 0)| \le L_s \qquad (7.4.9)$$

and

$$\lim_{s \to \infty} \frac{L_{s+1}}{L_s} = 0 . \qquad (7.4.10)$$

Polynomial Extrapolation

The stability polynomial for the Euler scheme is given by

$$\mu_E(z) = 1 + z . \qquad (7.4.11)$$

Hence, one can readily deduce that the stability polynomial for an arbitrary element T_{r0} of the extrapolation Table 7.2 is

$$B_E(z, r) = (1 + \frac{z}{2^r})^{2^r} . \qquad (7.4.12)$$

Any diagonal element $T_{0,s}$ of the extrapolation table based on polynomial extrapolation satisfies the following recurrence relation:

$$T_{0,s} = \frac{2^s T_{r+1,s-1} - T_{r,s-1}}{2^s - 1} . \qquad (7.4.13)$$

The stability function $\mu_p(z, s)$ for $T_{0,s}$ given in (7.4.13) can be readily obtained as

$$\mu_p(z, s) = \sum_{r=0}^{s} \alpha_{s,s-r}(1 + \frac{z}{2^r})^{2^r} , \qquad (7.4.14)$$

with the coefficients $\alpha_{s,s-r}$ recursively given by

$$\alpha_{0,0} = 1 ,$$

$$\alpha_{s-1,-1} = 0 ,$$

$$\qquad (7.4.15)$$

$$\alpha_{s-1,s} = 0 ,$$

Table 7.8.

M	1	2	3	4	5	6
z	-2	-2.785	-4.23	-9.06	-10.88	-13.88

$$\alpha_{s,r} = \frac{2^s \alpha_{s-1,s-r} - \alpha_{s-1,s-1-r}}{2^s - 1} .$$

The region of absolute stability R for (7.4.13) is given by

$$R = \{z \mid |\mu_p(z, M)| < 1\} . \tag{7.4.16}$$

The intercept of R with the real line is illustrated in Table 7.8, and the stability regions are given in Figure 7.1.

The stability for the inverse Euler (7.2.14) is given by

$$\mu_I(z) = (1-z)^{-1} . \tag{7.4.17}$$

The r-th element, i.e. T_{r_0}, of the zero-th column extrapolation table, hence, has a stability function

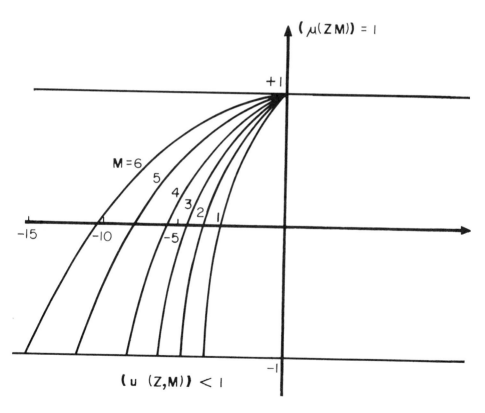

Figure 7.1: Stability Region for Euler (+) Polynomial Extrapolation

$$\mu_l(z, r) = (1 - \frac{z}{2^r})^{-2^r} . \tag{7.4.18}$$

If subsequent columns of the extrapolation table are generated by the polynomial approach, the stability function of a diagonal element $T_{0,s}$ can be obtained as

$$\beta_l(z, r) = \sum_{r=0}^{s} \alpha_{s,s-r} \mu_l(z, r) , \tag{7.4.19}$$

with coefficients $\alpha_{s,s-r}$ given by (7.4.15). It can be readily shown that

$$\lim_{z \to -\infty} |\beta_l(z, r)| = 0 ,$$

which indicates the A-stability of the inverse Euler combined with polynomial extrapolation. For scalar equations, nonlinear methods present theoretical difficulties for stability analysis.

7.5. Practical Implementation of Extrapolation Processes

The papers of Bader and Deuflhard (1983) and Deuflhard (1983, 1985) have greatly enhanced the competitiveness of the extrapolation schemes with the highly sophisticated codes of Adams-Moulton/backward differentiation methods, for both stiff and nonstiff initial value problems.

For theoretical and practical points of view, Aitken-Neville (1932, 1934) polynomial extrapolation applied to the Gragg modified midpoint rule (for nonstiff) and to the semi-implicit midpoint scheme (for stiff problems) takes precedence over the rational extrapolation scheme of Bulirsch and Stoer (1966a).

Each element $T_{r,s}$ of the extrapolation Table 7.2 denotes a special approximation to $y(x_{n+1}, h_n)$ with error

$$e_{1,r} = T_{1,r} - y(x_{n+1}, h_n) . \tag{7.5.1}$$

For any appropriate vector norm, we assume the Toeplitz condition

$$\frac{N_r}{N_{r+1}} \leq \alpha < 1 . \tag{7.5.2}$$

The local error can be approximated by the diagonal error criterion

$$e_{1,r} := T_{1,r} - T_{0,r+1} \leq \text{TOL} \tag{7.5.3}$$

for some user specified tolerance TOL.

Because $T_{0,r+1}$ is of higher order than $T_{1,r}$, for sufficiently small h,

$$e_{1,r+1} < e_{1,r} . \tag{7.5.4}$$

In the event that $r+1 \leq M$, the next integration meshsize is given as

$$h_{n+1} = [\frac{TOL}{e_{1,r}}]^{1/p_r} * h_n \tag{7.5.5}$$

with

$$p_r = \gamma r . \tag{7.5.6}$$

8. STIFF INITIAL VALUE PROBLEMS

8.1. The Concept of Stiffness

Consider the scalar initial value problem of Dahlquist et al. (1980):

$$y' = \lambda y + (\mu + i)e^{ix}, \; y(0) = y_0 \tag{8.1.1}$$

for some complex constant λ with $\text{Re}(\lambda) < 0$ and a real number μ; $i^2 = -1$.

The theoretical solution to (8.1.1) can be obtained as

$$y(x) = \frac{\mu + i}{-\lambda + i}e^{ix} + (y_0 - \frac{\mu + i}{-\lambda + i})e^{\lambda x}. \tag{8.1.2a}$$

As x increases, this rapidly approaches the steady state

$$y(x) = \frac{\mu + i}{-\lambda + i}e^{ix}. \tag{8.1.2b}$$

Definition 8.1. The transient phase of (8.1.1) is the interval $[0, x_\varepsilon]$ where the point x_ε is given as

$$e^{\lambda x_\varepsilon} = \text{TOL}, \tag{8.1.3a}$$

with TOL being the allowable error tolerance.

An alternative but over-simplified definition of the transient phase for (8.1.1) is

$$x_\varepsilon = -\frac{1}{\text{re}(\lambda)}. \tag{8.1.3b}$$

In case of the differential system

$$y' = Sy ,$$ (8.1.4)

(for some constant matrix S), the length of the transient phase is

$$x_\varepsilon = \min_{1 \leq i \leq k} - \frac{1}{re(\lambda_i)} ,$$ (8.1.3c)

where $k \leq m$ is the number of transient components. If, however, all the eigenvalues of (8.1.4) have negative real parts, then the component with the smallest real part does not have a transient phase because, as the dominant one, it never becomes insignificant.

In case of the non-homogeneous differential system

$$y' = Sy + g(x) ,$$ (8.1.5)

the forcing function $g(x)$ can re-excite the stiff components, and a transient phase can be re-entered.

In the transient phase, accuracy considerations compel a numerical integrator to adopt a meshsize of the order of the smallest time constant x_ε that is still significant; e.g., for parabolic equations, there is a sequence of different transient regions determined by the most negative eigenvalue. Outside the transient phase, there is always the desire to adopt a reasonably large meshsize, but this is often restricted by stability considerations.

Approximate solutions $\{y_n\}$ to (8.1.1) can be generated with the explicit Euler's scheme (3.2.2) as

$$y_{n+1} = y_n + h[\lambda y_n + (\mu + i)e^{ix_n}] .$$ (8.1.6)

The difference $\delta y(x)$ of two neighboring solutions, emanating from different initial conditions (often due to roundoff errors), satisfies the initial value problem

$$\frac{d}{dx} \delta y = \lambda \delta y , \; \delta y(0) = \eta ,$$ (8.1.7)

whose theoretical solution

$$\delta y(x) = e^{\lambda x} \eta$$ (8.1.8)

approaches zero rapidly.

The question then arises regarding the behavior of the difference δy_n at $x = x_n$ of two numerical solutions to (8.1.7), due to slightly different initial conditions.

We now assume two neighboring initial conditions, $y(0) = y_0$, $z(0) = z_0$ and set $\eta = y_0 - z_0$. One may then wish to know if there is any possible resemblance in the behavior of δy_n and $\delta y(x_n)$ for $n \geq 0$.

It can be shown that δy_n satisfies the following recurrence relation:

$$\delta y_{n+1} = (1 + h_n \lambda) \delta y_n ,$$ (8.1.9)

which possess the properties

$$\lim_{n \to \infty} \delta y_n = 0 ,$$ (8.1.10)

provided

$$|1 + h_n \lambda| < 1 ,$$ (8.1.11)

and

$$\lim_{n \to \infty} \delta y_n = \infty ,$$ (8.1.12)

if

$$|1 + h_n \lambda| > 1 .$$ (8.1.13)

An attempt to fulfill Condition (8.1.11) when solving a stiff system with the explicit Euler scheme will require the adoption of an intolerably small meshsize throughout the entire interval of integration. The resultant computational effort will be enormous if $|b - a| \gg 1/|\lambda|$.

Stiff initial value problems were first encountered in the study of the motion of springs of varying stiffness, from which the problem derives its name.

Stiff initial value problems are of frequent occurrence in the mathematical formulation of physical situations in control theory and mass action kinetics, where processes with widely varying time constants are usually encountered.

Historically, two chemical engineers (Curtis and Hirschfelder 1952) proposed the first set of numerical integration formulas that are well-suited to stiff initial value problems. They suggested the adoption of

$$y_{n+1} - y_n = h f_{n+1}$$ (8.1.14)

and

$$y_{n+2} - \frac{4}{3} y_{n+1} + \frac{1}{3} y_n = \frac{2}{3} f_{n+2} ,$$ (8.1.15)

of error orders one and two, respectively.

Both schemes are implicit and belong to the well known class of backward differentiation formulas (BDF), (8.1.14) being the implicit Euler's scheme, whose implementation is discussed in Section 8.3.

The application of (8.1.14) to (8.1.1) leads to the difference equation

$$y_{n+1} - y_n = h[\lambda y_{n+1} + [\mu + i) e^{i x_{n+1}}] ,$$ (8.1.16)

whose error propagation is described by the difference equation

$$\delta y_{n+1} = (1 - h_n \lambda)^{-1} \delta y_n .$$ (8.1.17)

Equation (8.1.17) satisfies Relation (8.1.11) for all $h_n > 0$ and is A-stable (see Definition 3.9). Identical results hold for Formula (8.1.15). The Dahlquist Barrier Theorem (1963) imposes a severe limitation on the attainable order of A-stable linear multistep methods.

Theorem 8.1 (Dahlquist 1963). *An A-stable linear multistep method*

(i)   must be implicit, and

(ii) the most accurate A-stable LMM is the trapezoidal scheme (3.4.2) of order
   $p = 2$ and error constant $c_3 = -1/12$.

Implicitness introduces the problem of solving nonlinear algebraic systems using a modified Newton iteration at every interval or at certain intervals (cf. Section 8.3).

The Dahlquist Barrier Theorem can be circumvented by adopting unconventional numerical integrators, some of which are

(i)   nonlinear multistep schemes (Lambert 1973a,b; Lee 1974, 1976; Lee and Preiser 1977),

(ii) multiderivative multistep formulas (Enright 1974a,b; Fatunla 1978a, 1980; Cash 1981a; Brown 1974; Addison 1979; Sacks-Davis and Shampine 1981),

(iii) exponential fitting (Liniger 1969; Brunner 1972; Lee 1974, 1976; Liniger and Willoughby 1970; Iserles 1978a, 1979a, 1981a; Iserles and Powell 1980; Ehle 1968, 1973; Ehle and Picel 1975; Norsett 1975, 1978, Cash 1981c,d),

(iv) extrapolation processes (Lindberg 1973; Bader and Deuflhard 1983; Deuflhard 1983, 1985; Shampine 1983c), and

(v)   cyclic linear multistep methods (Tendler et al. 1978; Bickart et al. 1971; Bickart and Picel 1973; Albrecht 1978a,b, 1985, 1986, 1987; Tischer and Sacks-Davis 1983; Tischer and Gupta 1983a,b, 1985; Iserles 1982, 1984; Cash 1983a,b; Rosser 1967; Chu and Hamilton 1987; Zhou 1986).

The Dahlquist (1963) barrier theorem can also be bypassed by implicit R-K methods, by methods that exploit the Jacobian (i.e., modified Adams methods of Norsett and Jain, variable coefficient R-K methods of Van der Houwen), or by using the Lawson (1967a) transformation.

Singularly perturbed problems are special cases of stiff differential systems. For example, we consider the 2×2 linear system

$$\begin{bmatrix} y_1 \\ y_2 \end{bmatrix}' = \begin{bmatrix} a_{11} & a_{12} \\ a_{21} & a_{22} \end{bmatrix} \begin{bmatrix} y_1 \\ y_2 \end{bmatrix}, \tag{8.1.18}$$

whose eigenvalues λ_1, λ_2 are such that $0 > \lambda_2 > \lambda_1$. The System (8.1.18) is equivalent to the following second order differential equation:

$$y'' - (a_{11} + a_{22})y' + (a_{11}a_{22} - a_{12}a_{21})y = 0. \tag{8.1.19}$$

Let λ_1, λ_2 be the roots of the characteristic polynomial

$$\lambda^2 - (a_{11} + a_{22})\lambda + (a_{11}a_{22} - a_{12}a_{21}) = 0. \tag{8.1.20}$$

The following algebraic relations hold:

$$\lambda_1 + \lambda_2 = a_{11} + a_{22}$$

$$\tag{8.1.21}$$

$$\lambda_1 \lambda_2 = a_{11}a_{22} - a_{12}a_{21}.$$

The System (8.1.18) can then be recast as a singular perturbation problem

$$\varepsilon y'' - (\frac{\lambda_2}{\lambda_1} + 1)y' + \lambda_2 y = 0, \tag{8.1.22}$$

with $\varepsilon = 1/\lambda_1 \to 0$.

The concept of stiff initial value problems can be better appreciated by considering the following general linear systems with constant coefficients:

$$y' = Ay + d(x), \ y(a) = y_0, \tag{8.1.23}$$

where A is an $m \times m$ matrix with real entries and $B(x)$, y, y' are m-vectors.

The theoretical solution to (8.1.23) was obtained in (1.4.7) as

$$y(x) = \sum_{i=1}^{m} \alpha_i e^{\lambda_i x} c_i + y_p(x), \tag{8.1.24}$$

where λ_i, $i = 1(1)m$ are the eigenvalues of A, with c_i, $i = 1(1)m$ the corresponding eigenvectors. $y_p(x)$ is a particular solution to (8.1.23), and α_i, $i = 1(1)m$ are real coefficients that are uniquely specified by the associated initial conditions $y(a) = y_0$.

Definition 8.2 (Enright 1983a). The stiffness ratio S of the system (8.1.23) is given as

$$S = \max_i \left\{ \frac{|\lambda_i|(b - a)}{\ln(\text{TOL})} \right\}, \tag{8.1.25}$$

where $\ln(\text{TOL})$ is the exponential logarithm of TOL.

The stiffness ratio as given by (8.1.25) is a measure of the dispersion of the time constants for (8.1.23), and, in real problems, may be of the order of 10^{+8}.

Definition 8.3. The initial value problem (8.1.23) is said to be stiff if it satisfies (8.1.29a) and (8.1.29b); i.e., if

(i) $\text{re}(\lambda_i) < 0$, $i = 1(1)m$, and

(ii) the stiffness ratio $S \gg 1$.

However, it should be noted that this is a rather imprecise definition mathematically. Stiffness occurs if the stepsize is restricted by stability, rather than order, considerations.

If matrix $A = A(x)$ (i.e., has variable coefficients), the first condition in Definition 8.3 is no longer binding, and this is considered in the stiff stability criteria (see Section 8.5, Definition 8.12).

The numerical solution to (8.1.23) generated with the explicit Euler's scheme is given as

$$y_0 = y(a) , \quad \frac{(y_{n+1} - y_n)}{h_n} = Ay_n , \quad h_n = x_{n+1} - x_n . \tag{8.1.26}$$

It is expected that the difference quotient $(y(x_{n+1}) - y(x_n))/h$ be closely approximated by (8.1.26) in order to adopt meshsize $h_n = h$. This is possible provided h_n is chosen to be smaller than the smallest time constant; otherwise, the propagation of the perturbation in (8.1.26) will not be equivalent to those in (8.1.23).

This point is well made if one considers the application of any explicit Runge-Kutta process to the error differential system of (8.1.23), which is given as

$$\delta\prime = A\delta , \quad \delta(a) = \eta . \tag{8.1.27}$$

This yields the difference equation

$$\delta_{n+1} = P(z)\delta_n , \quad z = \lambda h_n , \tag{8.1.28}$$

where P is a polynomial whose degree is less than or equal to the number of stages of the R-K scheme (e.g., the Fehlberg (4,5) scheme has six stages, and P is of degree six).

Gaffney (1984) discussed the concept of stiff oscillatory systems.

Definition 8.4 (Gaffney 1984). The IVP (2.1.4) is considered to be stiff oscillatory if the eigenvalues $\{\lambda_j = u_j + iv_j, \ j = 1(1)m\}$ of the Jacobian $J = (\partial f / \partial y)$ possess the following properties:

$$u_j < 0, j = 1(1)m , \tag{8.1.29a}$$

$$\max_{1 \le j \le m} |u_j| > \min_{1 \le j \le m} |u_j| , \tag{8.1.29b}$$

or if the stiffness ratio satisfies

$$S = \max_{i,j} |\frac{u_i}{u_j}| > 1$$

and

$$|u_j| < |v_j| \tag{8.1.29c}$$

for at least one pair of j in $1 \le j \le m$.

8.2. Stiff and Nonstiff Algorithms

Nonstiff algorithms have a finite region of absolute stability (RAS), while stiff algorithms have unbounded RAS. This accounts for why stiff algorithms accommodate the use of a large meshsize outside the transient (nonstiff) phase.

In the transient phase (boundary layer), automatic codes attempt to identify the optimum meshsize that keeps the local truncation error within the limit of the requested accuracy.

Outside the transient phase (i.e., in the stiff phase) the growth of propagated errors (instability) controls the choice of meshsize if a nonstiff method is adopted, because stability restrictions are independent of the accuracy requirements. This point will be illustrated with the moderately stiff system (Shampine 1975):

$$y' = \begin{bmatrix} -0.1 & -199.9 \\ 0 & -200 \end{bmatrix} y , \; y(0) = (2, 1)^T \tag{8.2.1}$$

in the interval $[0, 50]$. The theoretical solution to (8.2.1) is

$$y(x) = \begin{bmatrix} \exp(-0.1x) + \exp(-200x) \\ \exp(-200x) \end{bmatrix} , \tag{8.2.2}$$

and its transient phase is $[0, x_\varepsilon]$ with the point x_ε specified as

$$x_\varepsilon = \{x: \; |y_2(x)| = TOL\}. \tag{8.2.3}$$

Table 8.1a gives the error for the nonstiff method of Fehlberg RKF (4,5) scheme in a subinterval of the stiff phase $1 \le x \le 50$ for various tolerances $\{TOL = 10^{-q}, q = 1(1)12\}$.

Both the cost (number of function evaluations) and the average meshsizes in Table 8.1a are independent of the accuracy requirements if h_n are effectively chosen as to ensure that λh (average h) always fall on the boundary of the RAS.

Shampine (1975, 1980) also highlighted in Table 8.1b the disparity in the efficiency of adopting a nonstiff code (STEP) based on the Adams-Moulton method (cf. Shampine and Gordon 1975) and of adopting a stiff code (DIFSUB) based on the BDF (cf. Gear 1971a,b). Both codes were applied to the IVP (8.2.1) in the boundary layer $[0, x_\varepsilon]$ and in the stiff zone $[x_\varepsilon, 50]$.

In the transient phase, stability does not pose any serious problem, but the stepsize is chosen so as to resolve the rapid change in order to cope with accuracy requirements. The nonstiff code is more efficient than the stiff code, which is more expensive per integration step in this zone. However, in the stiff zone $[x_\varepsilon, 50]$ where stability dominates, the stiff code exhibits its superiority. The Adams code is independent of the allowable error tolerance, as in the case of the RKF45 (see Table 8.1a). The stiff code DIFSUB responds to the allowable error tolerance, as shown in Table 8.1b.

Table 8.1a. Performance of Nonstiff Method (RKF45)
on a Moderately Stiff Problem (8.2.1)
Index $a(-b) = a \times 10^{-b}$

q	Function Evaluations	Maximum Error	Average Meshsize
1	21672	3(-3)	0.01536
2	21744	3(-4)	0.01536
3	21671	3(-5)	0.01537
4	21697	3(-6)	0.01537
5	21725	3(-7)	0.01536
6	21687	3(-8)	0.01536
7	21694	3(-9)	0.01536
8	21680	3(-10)	0.01536
9	21747	3(-11)	0.01537
10	21613	3(-12)	0.01535
11	21691	3(-12)	0.01536
12	21727	3(-12)	0.01536

8.3. Solution of Nonlinear Equations and Estimation of Jacobians

Solving stiff and highly oscillatory IVPs (2.1.4) normally requires the adoption of implicit numerical integrators because these integrators possess better stability properties. The implicit schemes, however, require the solution of nonlinear equations of the form

$$F(y) = 0 , \qquad (8.3.1)$$

where F and y are m-vectors.

As an illustration, the application of the simplest implicit numerical integrator--the backward Euler scheme--to (2.1.4) yields the following first order difference equation:

$$y_{n+1} = y_n + hf(x_{n+1}, y_{n+1}) , \qquad (8.3.2)$$

which can be expressed in the Form (8.3.1) as follows:

$$F(y_{n+1}) = y_{n+1} - y_n - hf(x_{n+1}, y_{n+1}) = 0 . \qquad (8.3.3)$$

If a suitable starting value $y_{n+1}^{[0]}$ is available, (8.3.2) can be solved iteratively as

$$y_{n+1}^{[\mu+1]} = y_n + hf(x_{n+1}, y_{n+1}^{[\mu]}) , \mu = 0, 1, 2,.... \qquad (8.3.4)$$

Definition 8.5. An iteration formula is said to be of order p if there exist a finite constant c such that

Table 8.1b. Performance of Stiff (BDF) and Nonstiff (AMF)
Codes on Moderately Stiff Problem (Shampine 1975)
Index $a(-b) = a \times 10^{-b}$

TOL = 10^{-q}	Nonstiff Code STEP Adams-Moulton Shampine & Gordon (1975)			Stiff Code DIFSUB Backward Differentiation Gear (1971)		
	Function Evaluations		Maximum Error	Function Evaluations		Maximum Error
q	$[0,x_\varepsilon]$	$[x_\varepsilon,50]$		$[0,x_\varepsilon]$	$[x_\varepsilon,50]$	
1	13	13022	9(-1)	20	33	6(-2)
2	27	13071	9(-2)	48	43	1(-2)
3	45	12999	1(-2)	66	106	1(-3)
4	67	13068	1(-3)	96	125	1(-4)
5	87	12914	1(-4)	133	162	2(-5)
6	112	13307	1(-5)	190	147	2(-6)
7	141	13283	5(-7)	232	245	2(-7)
8	167	13190	6(-8)	344	213	4(-8)
9	205	13326	7(-9)	447	247	4(-9)
10	261	13441	4(-10)	575	304	6(-10)
11	294	14191	9(-11)	733	385	4(-11)
12	352	14734	6(-12)	992	464	7(-12)
13	426	15022	1(-12)	1298	607	4(-12)

$$\lim_{\substack{\mu \to \infty \\ y_{n+1}^{[\mu]} \to y(x_{n+1})}} \frac{y_{n+1}^{[\mu+1]} - y(x_{n+1})}{y_{n+1}^{[\mu]} - y(x_{n+1})^p} = c , \qquad (8.3.5)$$

where $y(x_{n+1})$ is the exact solution.

It can be shown that (8.3.4) converges linearly (i.e., $p = 1$) provided (8.3.3) satisfies the hypothesis of the contraction mapping theorem (see Wait 1979 and Kolmogovov and Fomin 1957).

Because the one point iteration scheme (8.3.4) converges only for very small values of h when the IVP is stiff (since f has a very large Lipschitz constant with respect to y), the Newton's scheme, which is known to have a faster convergence rate, is more commonly adopted. It is

$$y_{n+1}^{[\mu+1]} = y_{n+1}^{[\mu]} - G(y_{n+1}^{[\mu]})F(y_{n+1}^{[\mu]}) , \quad \mu = 0, 1, 1,..., \qquad (8.3.6a)$$

where $G(y_{n+1}^{[\mu]})$ is an $m \times m$ matrix specified as

$$G(y_{n+1}^{[\mu]}) = \frac{\partial F}{\partial y}(y_{n+1}^{[\mu]})^{-1} = (I_m - h\frac{\partial f}{\partial y}(x_{n+1}, y_{n+1}^{[\mu]})^{-1} , \qquad (8.3.6b)$$

where I_m is an identity matrix of order m.

Ortega and Rheinboldt (1970, page 312) give conditions that ensure the convergence and the rate of convergence of (8.3.6).

Newton's Scheme (8.3.6a) calls for a substantial additional computational effort at each integration step. This includes the evaluation of the Jacobian $(\partial f/\partial y)$ and the LU decomposition of the iterative matrix $(I_m - h\partial f/\partial y\,(x_{n+1}, y_{n+1}^{[\mu]}))$ at each iteration. This can be quite demanding, particularly in those practical problems of large dimension $m = O(10^3)$.

Ideally, numerical analysts minimize the computational effort with minimum risk to stability and convergence. Hence, variants of Newton's scheme are adopted in practice.

A generalization of (8.3.6) can be written in the form

$$W\delta_{n+1}^{[\mu+1]} = -F(y_{n+1}^{[\mu]}),\qquad\qquad(8.3.7)$$

where W is an $m \times m$ matrix, and $\delta_{n+1}^{[\mu+1]}$ is the correction vector given by

$$\delta_{n+1}^{[\mu+1]} = y_{n+1}^{[\mu+1]} - y_{n+1}^{[\mu]}.\qquad\qquad(8.3.8)$$

Some possible choices of W are

(a) $W = I_m$,   (8.3.9)

the $m \times m$ identity matrix. Hence, (8.3.7) reduces to the functional iteration scheme stated in (8.3.4), which is adequate only at a nonstiff integration step.

(b) $W = \dfrac{\partial F}{\partial y}(y_{n+1}^{[\mu]})$.   (8.3.10)

Hence (8.3.7) reverts to the Newton's scheme, given by Equation (8.3.6).

(c)   If (8.3.10) is replaced by

$$W = \frac{\partial F}{\partial y}(y_{n+1}^{[0]}),\qquad\qquad(8.3.11)$$

then we have a simplified Newton's scheme that reduces the number of LU decompositions to just one per integration step. The reduction in the computational effort could be quite substantial, particularly in a real life stiff system whose dimensions may be of the order of several thousand.

(d)   Still more economical than procedure (c) is the "oversimplified" Newton scheme, whereby (8.3.11) is replaced by

$$W = \frac{\partial F}{\partial y}(y_j^{[0]}),\, j < n+1.\qquad\qquad(8.3.12)$$

This is the approach commonly adopted in the solution of stiff systems, and the matrix W is only recomputed if there are convergence difficulties.

It must be emphasized that (c) and (d) give only linear convergence. Their importance is not in speeding up convergence (unlike the Newton-Raphson method), but in extending the basin of attraction of the solution of $F(y) = 0$.

Most recent codes like LSODES exploit certain properties like the sparsity of the Jacobian or the iteration matrix.

(e)   Quasi-Newton schemes completely eliminate the matrix decomposition in (8.3.7), as this could be rather expensive in terms of computing time and computer memory.

Instead, the matrix $G(y_{n+1}^{[s]})^{-1}$ is approximated by a matrix $H_{n+1}^{[s]}$, which has to be updated until an acceptable solution is realized. This approach is expensive, too. Interested readers can look up the details of this approach in textbooks dedicated to the solution of nonlinear equations, e.g., Ortega and Rheinboldt (1970), Alfeld (1986), Hindmarsh (1985), and Hindmarsh and Brown (1986).

Norsett and Thomsen (1986a,b) expressed the Rate Of Convergence (ROC) of the iterative scheme as

$$ROC = \frac{Y^{(s+1)} - Y^{(s)}}{Y^{(s)} - Y^{(s-1)}} . \tag{8.3.13}$$

$ROC < 1$ implies a fast convergence, while $ROC > 1$ indicates divergence. A good starting value $Y^{(0)}$ is highly desirable, and the Jacobian should be recomputed if convergence rate is too slow. In the code SIMPLE, a new stepsize is recomputed as

$$h_{new} = \frac{0.8}{ROC} \, h \; 0.9 \tag{8.3.14}$$

if $ROC > 0.8$.

Estimate of the Jacobian

The nonlinear IVP

$$y' = f(x, y), \; y(a) = y_0 \tag{8.3.15}$$

may be adequately represented by the linear system (8.1.23), using the variational equation

$$y' = J(x)(y - y(x)) + f(x, y(x)). \tag{8.3.16}$$

The Jacobian $J(x)$ can be regarded locally as a constant matrix, while the vector $d(x) \equiv f(x, y(x))$ is normally ignored in the stability analysis of numerical integration schemes. Actually, $d(x)$ drops out when considering linear systems, because stability refers to the way the difference of two solutions behaves.

The elements of the Jacobian

$$J(x) \equiv \frac{\partial f}{\partial y}(x_{n+k}, y_{n+k})$$

can be reasonably approximated by forward differences as follows

$$J_{ij} \equiv \frac{f_i(x, y+\delta_j e_j) - f_i(x, y)}{\delta_j} \; , \; 1 \le i \, , j \le m \; , \qquad (8.3.17)$$

where e_j is a unit vector

$$e_j = (0, 0,..., 0, 1, 0,..., 0) \; , \; 1 \le j \le m \; .$$

The increment δ_j is a real number and can be chosen as to minimize the l.t.e. as well as roundoff errors (Curtis and Reid 1974). A special structure (e.g., sparsity or bandedness) of the Jacobian can be exploited in order to reduce the computational effort in the generation of this approximation and in the solution of the resultant linear systems.

Further discussion on this topic can be found in Chapter 9.

8.4. Region of Absolute Stability

Many practical initial value problems (8.1.23) have the property that the matrix A in (8.1.23) has distinct eigenvalues. This ensures the existence of a transformation matrix H such that

$$\delta = HZ \; , \; H \text{ nonsingular} \; , \qquad (8.4.1)$$

which reduces (8.1.27) to the uncoupled system

$$z' = \Omega z \; , \qquad (8.4.2)$$

where $\Omega = \text{diag}(\lambda_1, \lambda_2,..., \lambda_m)$. The IVP (8.4.2) provides the basis for the choice of the scalar test problem

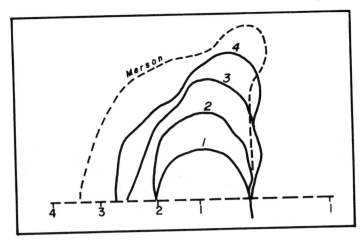

Figure 8.1:
Explicit Runge-Kutta, $p = 1(1)4$; Merson Five-Stage of Order Four

$$y' = \lambda y, \tag{8.4.3}$$

which is normally used for the stability analysis of numerical integration schemes. If the explicit k-stage Runge-Kutta scheme is applied to the test Equation (8.4.3), we get

$$y_{n+1} = P_k(z)y_n \, , \, z = h\lambda \, , \tag{8.4.4}$$

where P_k is a polynomial of degree $\leq k$. If $k \leq 4$, the order of the method can be $p = k$, and P_k coincides with the first $k+1$ terms in the power series of $\exp(z)$, as illustrated in Table 4.4. Hence, all explicit R-K schemes of order $p = k \leq 4$ have identical regions of absolute stability, as illustrated in Figure 8.1.

In the case of $k > p$, the RAS may change, as illustrated with Merson's five-stage scheme (4.4.13), whose order $p = 4$.

Some implicit R-K schemes are known to enjoy excellent stability properties, e.g., the unique s-stage scheme of order $2s$ given by (4.6.10). This is further discussed in Section 8.6.

The application of the conventional LMM, given as

$$\rho(E)y_n = h\sigma(E)f_n \, , \tag{8.4.5}$$

to the test Problem (8.4.3), leads to the stability polynomial

$$\Pi(r, z) = \rho(r) - z\sigma(r) \, , \, z = h\lambda \, , \tag{8.4.6}$$

and the boundary locus curve is found by setting

$$z = \frac{\rho(r)}{\sigma(r)} \, , \, r = e^{i\theta} \, , \, 0 \leq \theta \leq 2\pi \, . \tag{8.4.7}$$

This curve is normally symmetric about the real axis. The upper half is obtained for $0 \leq \theta \leq \pi$, and a mirror image of this through the real line completes the RAS. The RAS for both the linear multistep method (5.3.8) and those of Adams-Bashforth and Nystrom is illustrated in Figures 8.2—8.4.

Predictor-Corrector Methods

The RAS of the predictor-corrector pair depends on two factors:

(i) the LMM characterized by (ρ_1, σ_1), (ρ_2, σ_2) for the predictor and corrector, respectively, and

(ii) the mode of implementation, i.e., either $P(EC)^\mu$ mode or $P(EC)^\mu E$ mode.

The stability polynomial for the predictor-corrector scheme in the $P(EC)^\mu E$ mode ($\mu \geq 1$) was given by Chase (1962) as

$$\Pi_E(r, z) = \rho_2(r) - z\sigma_2(r) + M_\mu(z) \cdot [\rho_1(r) - z\sigma_1(r)] \, , \tag{8.4.8}$$

where

$$M_\mu(z) = (z\beta_k)^\mu \frac{1 - z\beta_k}{1 - (z\beta_k)^\mu} \; ; \qquad\qquad (8.4.9)$$

while that of the $P(EC)^\mu$ mode was obtained as

$$\Pi_c(r, z) = \beta_k r^k [\rho_2(r) - z\sigma_2(r)] + \qquad\qquad (8.4.10)$$

$$M_\mu(z) \cdot [\rho_1(r)\sigma_2(r) - \rho_2(r)\sigma_1(r)] \; .$$

It is obvious that if $|z\beta_k| < 1$, then (8.4.9) implies that

$$\lim_{\mu \to \infty} M_\mu(z) = 0 \; , \qquad\qquad (8.4.11)$$

and, hence, from (8.4.8) and (8.4.10),

$$\Pi_E(r, z) \to \Pi(r, z) \; ,$$

and

$$\Pi_c(r, z) \to \beta_k r^k \Pi(r, z) \; ,$$

where $\Pi(r, z)$ is the stability polynomial for the corrector.

This shows that the stability polynomial of the P-C formula in either mode is essentially that of the corrector formula (ρ_2, σ_2) if μ is sufficiently large and h small. This is true if and only if the RAS for the corrector contains no points with $|z\beta_k| > 1$. Unfortunately, most corrector RAS do contain such points, as can be observed from Figures 8.5 and 8.6.

Lambert (1971) established the following relationship between the two modes:

(a) The general PEC algorithm has a region of absolute/relative stability identical with a particular explicit LMM P^*. There is little motivation for this because of the small RAS of the explicit LMM.

(b) The RAS of the general $P(EC)^{\mu+1}$ algorithm is identical to that of some $P^*(EC)^\mu E$ methods.

While Hull and Creemer (1963) and Ralston (1965b) argue in favor of the $P(EC)^{\mu+1}$ scheme, Brown, Riley and Bennet (1965), Crane and Klopfenstein (1965), Krogh (1966), and Hall (1967) advocate the $P(EC)^\mu E$ mode.

This is illustrated with the Adams-Bashforth-Moulton for both modes (see Figures 8.5 and 8.6).

Multiderivative Methods

Multiderivative multistep methods are characterized by polynomials $(\rho, \sigma_i, i = 1(1)s)$ and specified as

$$\rho(E)y_n - \sum_{i=1}^s h^i \sigma_i(E) f_n^{(i-1)} \; , \qquad\qquad (8.4.12a)$$

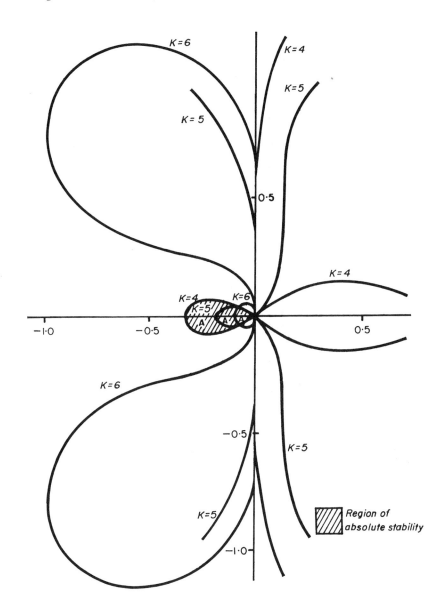

Figure 8.2: RAS of LMM (5.3.8)

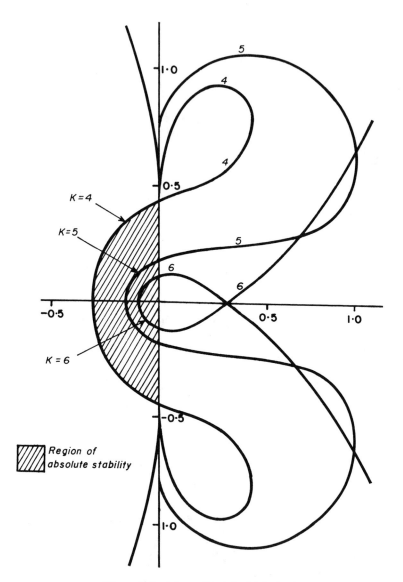

Figure 8.3: Adams-Bashford Method

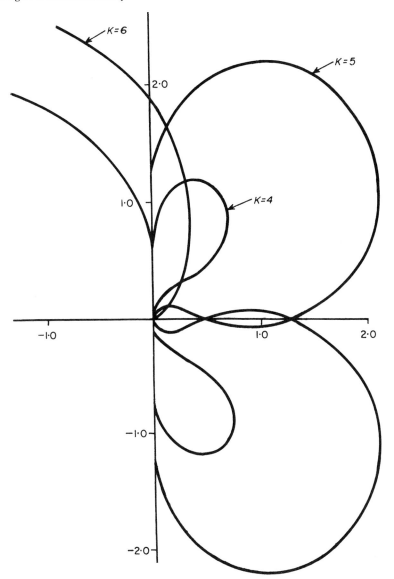

Figure 8.4: Nystrom Method

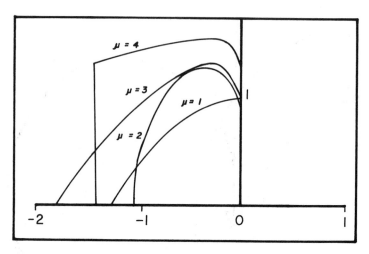

Figure 8.5: Adams-Bashforth-Moulton $P(EC)^{\mu}E$
Mode $\mu = 1(1)4$ order p

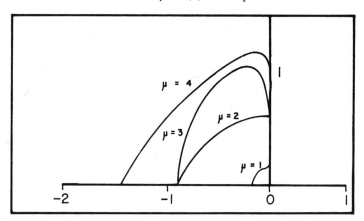

Figure 8.6: Adams-Bashforth-Moulton
$P(EC)^{\mu} \; \mu = 1(1)4$

where

$$\rho(E) = \sum_{j=0}^{k} \alpha_j E^j \qquad\qquad (8.4.12b)$$

and

$$\sigma_i(E) = \sum_{j=0}^{k} \beta_{ij} E^j \;, \; i = 1(1)s \; . \qquad\qquad (8.4.12c)$$

The stability polynomial for (8.4.12) can be obtained as

$$\Pi_s(r,\ z) = \rho(r) - \sum_{i=1}^{s} z^i \sigma_i(r) . \tag{8.4.13a}$$

Complications arise in the case $s > 1$, for if one attempts to use the boundary locus method, the boundary locus may consist of several separate curves. Daniel and Moore (1970) conjectured this. Wanner et al. (1978) used the order stars approach to establish some stability theorems.

Enright's second derivative formulas are particular cases of (8.4.12) where $s = 2$, with

$$\sigma_2(E) = \beta_{2k} E^k . \tag{8.4.13b}$$

The boundary locus for Enright's scheme is illustrated in Figure 8.7.

Brown (1974, 1977) proposed a special class of (8.4.12a) whereby

$$\sigma_i(E) = \beta_{ik} E^k ,\ i = 1(1)s ; \tag{8.4.14}$$

they are called Brown's $(k,\ s)$ methods. When $s = 1$, these methods are the conventional BDF (8.5.11) which have been shown to be Dahlquist-stable *iff* $1 \le k \le 6$ (Creedon and Miller 1975). Jeltsch and Kratz (1978) and Jeltsch

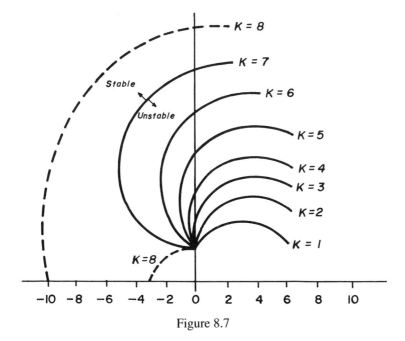

Figure 8.7

(1976, 1979) established that all Brown's (k, s) methods (8.4.12a, 8.4.14) are stable provided that $k \le 2(s+1)$.

Miller (1971) used the concept of Schur polynomials to determine RAS for LMM by obtaining those points in the z-plane where the stability polynomial is a Schur polynomial, i.e., when all the roots of $\Pi(r, s)$ lie within the unit circle.

Lambert (1973a, 1979) used this approach to investigate the RAS of numerical integrators for ODEs.

This procedure can be applied to any of the stability polynomials discussed in this section. A brief account of Miller's discussion is contained in Section 1.5.

8.5. Stability Criteria for Stiff Methods

We shall discuss some of the desirable stability properties for stiff methods.

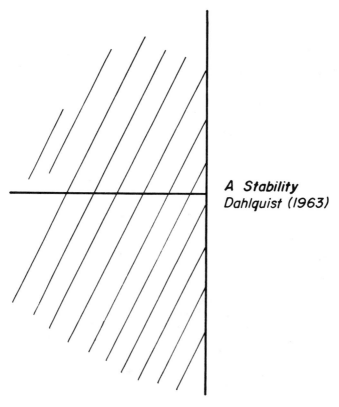

A Stability
Dahlquist (1963)

Figure 8.8

Definition 7.6 (Dahlquist 1963). A numerical integrator is said to be A-stable if its region of absolute stability R incorporates the entire left half of the complex plane denoted by C, i.e.,

$$R = \{z \; \varepsilon \; C \mid \mathrm{re}(z) < 0\} \; .$$

The concept of A-stability was originally formulated for LMM, but, because of Dahlquist's Barrier Theorem (1963), emphasis shifted to the derivation of implicit and semi-implicit one-step methods or related schemes that possess this most desirable property and even stronger stability properties. Some of these criteria are investigated in the works of Ehle (1968), Axelsson (1969), Chipman (1971), Alexander (1977, 1986), and Norsett (1974b).

The implicit and semi-implicit R-K processes demand large amounts of computation per step, and, besides, many one-step numerical integrators possess the undesirable property that when they are applied to the scalar test problem $y\prime = \lambda y$ with $\mathrm{re}(\lambda) < 0$, the resultant first order difference equation

$$y_{n+1} = R(z)y_n \qquad (8.5.1)$$

is such that the stability function $R(z)$ satisfies

$$\lim_{z \to -\infty} R(z) = \pm 1 \; . \qquad (8.5.2)$$

Gourlay (1970) observed this fault with the trapezium scheme (3.4.2) and suggested the adoption of the implicit midpoint rule

$$y_{n+1} = y_n + hf\left(x_{n+1/2}, \; \frac{1}{2}(y_n + y_{n+1})\right) \qquad (8.5.3)$$

for time-dependent problems. Equation (8.5.3) is a special case of the one-leg multistep scheme discussed in Dahlquist (1976, 1978a) and Nevanlinna and Liniger (1978, 1979). The general one-leg multistep scheme is described as

$$\rho(E)y_n = hf(\sigma(E)x_n, \; \sigma(E)y_n), \; \sigma(1) = 1 \; ; \; x_n = a + \sum_{r=0}^{n-1} h_r \; . \qquad (8.5.4)$$

Definition 8.7 (Dahlquist 1976). Contractivity implies that if the IVP (8.3.13) satisfies the relation

$$\mathrm{Re} < f(x, y) - f(x, z), \; y - z > \; \le 0 \qquad (8.5.5)$$

($<\cdot>$ a suitable inner product), then two neighboring numerical solutions $\{y_n\}$, $\{z_n\}$ must fulfill the following inequality:

$$z_n - y_n \le z_{n-1} - y_{n-1} \qquad (8.5.6)$$

for the corresponding norm $\cdot$. Contractivity implies A-stability.

Dahlquist (1976, 1978a, 1979a,b, 1981, 1983a), Odeh and Liniger (1975), Nevanlinna and Liniger (1978, 1979), and Sand (1981) contributed to the development and the analysis of the linear multistep method

$$\rho(E)y_n - h\sigma(E)f_n \qquad (8.5.7)$$

and its corresponding one-leg twin (8.5.4) for the numerical solution of stiff nonlinear differential system (8.3.13). Further discussion of contractive methods can be found in Sections 8.7 and 10.4.

Another stability property that is stronger than A-stability is the concept of L-stability, normally associated with one-step methods.

Definition 8.8. A one-step numerical integrator is said to be L-stable if

(i)   it is A-stable, and

(ii)   its resultant stability equation (8.5.1) satisfies the relation

$$\lim_{z \to \infty} |R(z)| = 0. \qquad (8.5.8)$$

Lambert (1980, pp. 32-34) showed that the stability region for L-stable schemes must encroach into the positive half of the complex plane C.

Another stability concept known to be stronger than A-stability and confined to one-step schemes is that of S-stability.

Definition 8.9 (Prothero and Robinson 1974). A one-step scheme is said to be S-stable if, when applied to the non-autonomous test equation

$$y' = \lambda(y - g(x)) + g'(x) \qquad (8.5.9)$$

(where $g(x)$ is any bounded function), yields a sequence $\{y_n, n \geq 0\}$ with the property

$$\frac{y_{n+1} - g(x_{n+1})}{y_n - g(x_n)} < 1, \qquad (8.5.10)$$

with $\forall h > h_0 > 0$ and $\forall \lambda$ satisfying $\text{Re}(\lambda) < -\lambda_0$.

S-stability => A-stability, and the converse does not hold.

Butcher (1975) introduced the concept of B-stability for one-step methods. It applies only to (dissipative) nonlinear stiff systems. This property has since been demonstrated for important classes of implicit R-K methods (Burrage 1978a,b; Burrage and Butcher 1979, 1981a; and by Wanner 1977). See Section 8.6 for further discussion on stronger stability properties that are associated with implicit/semi-implicit R-K methods.

The A-stability criteria is rather too restrictive as regards the attainable order of stiff LMM. This necessitates the introduction of more generous stability criterion, which are easily attainable by linear multistep methods of order $p \geq 3$.

Definition 8.10 (Widlund 1967). A numerical integration scheme is said to be $A(\alpha)$-stable for some $\alpha \in [0, \pi/2)$ if the wedge

$$S_\alpha = \{z: |\text{Arg}(-z)| < \alpha, z \neq 0\}$$

is contained in its region of absolute stability. The largest α (α_{max}) is called the angle of absolute stability on the argument of stability.

Definition 8.11. A scheme is $A(0)$-stable if it is $A(\alpha)$-stable for some $\alpha \, \varepsilon (0, \pi/2)$.

$$A(\frac{\pi}{2})\text{-stability} \equiv A\text{-stability} .$$

Kong (1977) and Grigorieff (1983) have shown that there exist $A(\alpha)$-stable k-step methods of order k for any $\alpha < \pi/2$ and any k.

Definition 8.12 (Cryer 1973). An integration procedure is A_0-stable if its region of absolute stability contains the negative real axis, i.e.,

$$\{z : \text{Im}(z) = 0 , \text{Re}(z) < 0\} .$$

It is obvious from the above definitions that

(i) $A(\pi/2)$-stability $<=>$ A-stability, and

(ii) L-stability $=> A(\alpha)$-stability $=> A(0)$-stability $=> A_0$-stability.

The previous definitions are concerned exclusively with stability properties, without regard for accuracy of the result generated. Gear (1969, 1971b) went a step further by incorporating accuracy in his concept of stiff stability.

Definition 8.13 (Gear 1969, 1971b). A numerical integrator is said to be stiffly stable provided it is

(a)   accurate in region R_2 specified as

$$R_2 = \{-D < \text{Re}(z) < \delta , |Im(z)| < \theta\} , D > 0 , \delta > 0 ,$$

and

(b)   absolutely stable in the region R_1

$$R_1 = \{\text{Re}(z) \leq -D\} .$$

Stiff stability $=> A(\alpha)$-stability for $\alpha = \tan^{-1}(\theta/D)$.

The rationale for stiff-stability is given by Gear (1971b, pp. 218-219).

Stiffly stable methods accommodate the existence of eigenvalues with small positive real parts $0 < \text{Re}(z) < \delta$, which may possibly occur locally in certain nonlinear IVPs.

Gear (1969) supported the adoption of the backward differentiation formulas (BDF)

$$\sum_{j=0}^{k}\alpha_j y_{n+j} = h\beta_k f_{n+k} , \alpha_k = 1 , k \geq 1 . \tag{8.5.11}$$

The coefficients are determined so as to ensure maximum possible order $p = k$. The resultant integration formulas are

(i) A-stable for $1 \leq k \leq 2$, and

(ii) stiffly stable for $3 \leq k \leq 6$.

In practice, stiffly stable schemes of order $p \leq 5$ are normally adopted, as the angle of absolute stability is too small for BDFs whose order p is greater than 5 (see Table 8.2).

The BDF is easily the most popular, most reliable, and best-tested stiff algorithm. Various versions of this approach have been proposed by Gear (1969, 1971b), Brayton et al. (1972), Hindmarsh (1974), Byrne and Hindmarsh in EPISODE (1975b), Hindmarsh in LSODE (1980), Shampine and Watts in DEPAC, a variant of LSODE (1979), FACSIMILE in Curtis (1980), and Gear in the divided difference approach (1980c) (cf. Table 9.4).

Enright et al. (1975) observed that the BDF are unrivalled for solving stiff problems whose eigenvalues are not close to the imaginary axis. For problems whose eigenvalues are close to the imaginary axis, multiderivative methods, cyclic multistep methods, and the implicit/semi-implicit R-K methods generally perform better (cf. Gaffney 1984).

The development of the BDF (8.5.11) is based on the numerical differentiation approach (cf. Section 5.3), and the associated stability polynomial for the BDF is given by

$$\Pi(r, z) = \rho(r) - z\beta_k r^k , \qquad (8.5.12)$$

which is very stable at infinity, as all the zeros of $\Pi(r, z)$ lie at the origin at $z = -\infty$. This is analogous to L-stability.

Gupta (1976) proposed the $A(\alpha, D)$-stability criterion, which incorporates both the properties of $A(\alpha)$-stability and those of stiff stability.

Definition 8.14 (Gupta 1976). An integration scheme is said to be $A(\alpha, D)$-stable if all the numerical solutions to the scalar test Problem (8.4.3) converge to zero as $n \to \infty$, with h fixed for $|\arg(-z)| < \alpha, D \leq \mathrm{Re}(z) < 0, |\lambda| \neq 0$ and for all $\mathrm{Re}(z) \leq D$.

Gupta proposed $A(\alpha, D)$-stable formulas of orders up to nine.

The angle of absolute stability α serves as a good indicator of the problem class for which a method is suitable—namely, those problems for which the eigenvalues of the Jacobian lie inside the wedge S_α of Figure 8.9. Hence, in an ODE solver, α can be used as an extra parameter to select the $A(\alpha)$-stable scheme which is most appropriate among a family of order k methods. The value of α can be supplied by the user or chosen automatically by some device in the code.

Definition 8.15 (Odeh and Liniger 1975). A numerical integration scheme is A_∞-stable if its region of absolute stability contains a neighborhood of ∞.

Dahlquist (1956) and Stetter (1973, p. 205) showed that a k-step method is A_∞-stable if and only if it satisfies the root condition.

Table 8.2. Angle of Absolute Stability of Some Popular Stiff Methods

ORDER	ODIOUS Tischer & Gupta 1983	STRIDE Burrage 1978	SDRROOT Tischer & Sacks-Davis 1983	BLEND Skeel & Kong 1977	MIRK Cash & Singhal 1982	STINT
1		90			90	90
2	90	90	89.99	90	90	90
3	90	[89.38,90]	84.88	90	90	89.43
4	90	[88.24,90]	70.63	90	90	80.88
5	86.64	[87.09,90]	44.70	89.4		77.48
6	76.32	[85.37,90]		87.0		63.25
7	57.66	[85.94,89.95]		82.9		33.53

ORDER	SDIRK(SIMPLE) Norsett & Thomsen 1984, 1986b	DIRK1 Alexander 1977 & Norsett 1974	BDF Gear 1971 & Norsett 1969	DSTIFF Gupta 1985	DIRK2 Cooper & Sayfy 1979
1			90	90	90
2		90	90	90	90
3	90	90	88.45	86.46	90
4		90	73.23	80.13	90
5			51.83	73.58	90
6			18.78	67.77	90
7				65.53	

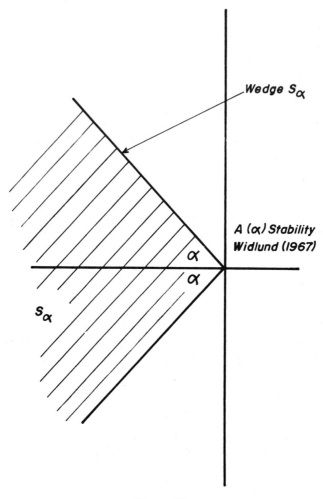

Figure 8.9

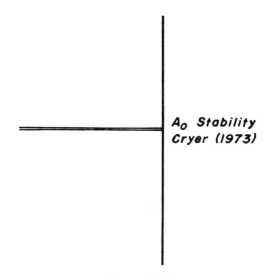

Figure 8.10

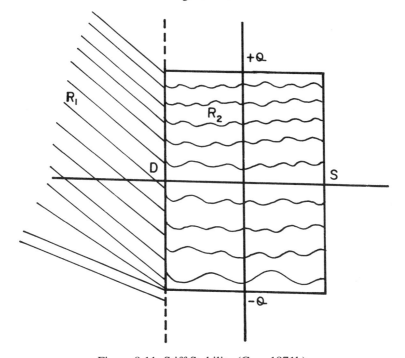

Figure 8.11: Stiff Stability (Gear 1971b)

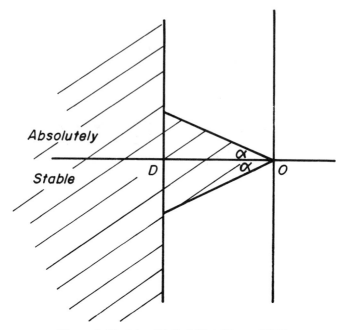

Figure 8.12: $A(\alpha, D)$-Stability (Gupta 1976)

Stetter (1973, pp. 211-213), among others, established that a k-step linear multistep method is convergent if and only if it is stable and consistent.

Although Widlund (1967) and Cryer (1973) established the equivalence of Dahlquist's (1963) A-stability barrier theorem for $A(\alpha)$-stability and A_0-stability, respectively, the barrier theorem is no longer true when A-stability is relaxed.

8.6. Stronger Stability Properties of IRK Processes

Dahlquist (1963) formulated the concept of A-stability for linear multistep methods. This criterion has since been extended to other classes of methods, particularly the Runge-Kutta methods, some of which are of high order and yet A-stable. Stronger stability criteria, including B-stability, AN-stability, BN-stability, and algebraic stability, have been formulated for the implicit Runge-Kutta methods

$$\begin{bmatrix} a_{11} & a_{1s} & c_1 \\ & & \\ & & \\ a_{s1} & a_{ss} & c_s \\ b_1 & b_s & 0 \end{bmatrix} \quad \text{or} \quad \begin{bmatrix} A & c \\ b^T & \end{bmatrix}, \qquad (8.6.1)$$

with A being a full matrix and $c_i = \sum_{j=1}^{s} a_{ij}$, $i = 1(1)s$.

Dahlquist (1963) defined A-stability with respect to the scalar autonomous system

$$y' = \lambda y \ , \ \text{re}(\lambda) < 0 \ . \tag{8.6.2}$$

Butcher (1975) introduced the concept of B-stability w.r.t., the more general test problem

$$y' = f(y(x)) \ , \ f : R^m \to R^m \ , \tag{8.6.3}$$

with the assumption that

$$<f(y) - f(z) \ , \ y - z > \ \leq 0 \tag{8.6.4}$$

for all y, $z \in R^m$, $<\cdot>$ being an inner product on R^m, and $\cdot$ the corresponding norm. AN-stability is associated with the test problem

$$y' = \lambda(x)y \ , \tag{8.6.5}$$

where $\lambda(x)$ takes values in the negative complex plane. BN-stability is associated with the non-autonomous test equation

$$y' = f(x, y(x)) \ , \ f : R^{m+1} \to R^m \ . \tag{8.6.6}$$

Let $\{y_n\}$, $\{z_n\}$ constitute two sequences of approximations computed with an Implicit Runge-Kutta (IRK) scheme using meshsize h.

Definition 8.16 (Butcher 1975). The method (8.6.1) is B-stable if, when it is applied to (8.6.3), we have that

$$|y_n - z_n| \leq |y_{n-1} - z_{n-1}| \tag{8.6.7}$$

whenever (8.6.4) holds.

Butcher further established the possibility of deriving a B-stable IRK of order $p = 2s - 2$, as well as the following sufficient condition for B-stability:

Theorem 8.2. *If (8.6.1) is such that $b_i \geq 0$, $i = 1(1)s$, A is nonsingular, the quadratic form*

$$\bar{Q}(\xi_1, ..., \xi_s) = \sum_{i,j=1}^{s} \bar{m}_{ij} \xi_i \xi_j \tag{8.6.8}$$

is non-negative where

$$\bar{m}_{ij} = b_i a_{ij}^{(-1)} + b_j a_{ji}^{(-1)} - \sum_{k=1}^{s} b_k a_{ki}^{(-1)} \sum_{k=1}^{s} b_k a_{kj}^{(-1)} \tag{8.6.9}$$

for $i,j = 1(1)s$, and $a_{ij}^{(-1)}$ is the (i,j)-th element of A^{-1}, then (8.6.1) is necessarily B-stable.

Burrage and Butcher (1979) formulated a simple algebraic criterion which gives a sufficient condition for B-stability. They also introduced a stronger concept called BN-stability.

Definition 8.17. Method (8.6.1) is said to be BN-stable if, when it is applied to (8.6.6) satisfying (8.6.4), then (8.6.7) holds.

BN-stability => A-, AN-, and B-stability for the linear non-autonomous differential systems.

The stability properties of (8.6.1) can be obtained by application of (8.6.1) to a single homogeneous linear equation (8.6.5), $\lambda: R \to C$ continuous. With meshsize $h = x_{n+1} - x_n$, we write

$$\xi_i = h\lambda(x_n + hc_i) , \quad i = 1(1)s \tag{8.6.10}$$

and set

$$\xi = \text{diag}(\xi_1, \xi_2, ..., \xi_s) . \tag{8.6.11}$$

It can be shown that if $z = \lambda h$, with λ constant, the numerical solution y_{n+1} satisfies the difference equation

$$y_{n+1} = R(z)y_n , \tag{8.6.12}$$

where

$$R(z) = 1 + zb^T(I - zA)^{-1}e \tag{8.6.13}$$

and

$$e = (1, 1,..., 1)^T . \tag{8.6.14}$$

In the case of (8.6.5), we obtain

$$y_{n+1} = K(\xi)y_n , \tag{8.6.15}$$

with $K(\xi)$ given as

$$K(\xi) = 1 + b^T\xi(I - A\xi)^{-1}e . \tag{8.6.16}$$

In general, $K(\xi)$ is a rational function in each of $\xi_1, \xi_2, ..., \xi_s$.

Definition 8.18. The method (8.6.1) is said to be AN-stable if its K function satisfies

$$|K(\xi)| \le 1 , \tag{8.6.17}$$

$\forall \xi = \text{diag}(\xi_1, ..., \xi_s)$ such that $\xi_i = \xi_j$ whenever $c_i = c_j$, and with $\text{Re}(\xi_i) \le 0$ for $i = 1(1)s$.

AN-stability implies A-stability as $R(z) = K(ze)$, but the converse does not hold. For the IRK scheme

$$\begin{pmatrix} \dfrac{1}{8} & \dfrac{1}{8} & \dfrac{1}{4} \\[2mm] \dfrac{3}{8} & \dfrac{3}{8} & \dfrac{3}{4} \\[2mm] \dfrac{1}{2} & \dfrac{1}{2} & 0 \end{pmatrix}, \tag{8.6.18}$$

we have

$$R(z) = \frac{2+z}{2-z} \tag{8.6.19}$$

and

$$K(\xi) = \frac{8 + 3\xi_1 + \xi_2}{8 - \xi_1 - 3\xi_2}. \tag{8.6.20}$$

The method (8.6.16) is A-stable, but is not AN-stable, as K is unbounded for $\mathrm{Re}(\xi_1), \mathrm{Re}(\xi_2) \le 0$.

Let $M = (m_{ij})$, $i,j = 1(1)s$ be the symmetric matrix whose elements m_{ij} are obtained from the coefficients of method (8.6.1) as

$$m_{ij} = b_i a_{ij} + b_j a_{ji} - b_i b_j. \tag{8.6.21}$$

Definition 8.19. Method (8.6.1) is BN-stable if, for two solution sequences $\{y_n\}$ and $\{z_n\}$ arising from its application to (8.6.6) (inherently stable), then

$$y_n - z_n \le y_{n-1} - z_{n-1}, \tag{8.6.22}$$

Theorem 8.3. *The method (8.6.1) is BN-stable if*

(i) $b_i \ge 0$, $i = 1(1)s$, *and*

(ii) *the quadratic form*

$$Q(\xi_1, \xi_2, ..., \xi_s) = \sum_{i,j=1}^{s} m_{ij} \xi_i \xi_j \tag{8.6.23}$$

is non-negative.

Proof. Let $v_0 = y_n - z_n$, $v_j = Y_j - Z_j$, $w_j = hf(x_n + hc_j, Y_j) - hf(x_n + hc_j, Z_j)$ for $j = 1(1)s$ and $v = y_{n+1} - z_{n+1}$.

We now have

$$v_i = v_0 + \sum_{j=1}^{s} a_{ij} w_j, \quad i = 1(1)s \tag{8.6.24}$$

and

$$v = v_0 + \sum_{j=1}^{s} b_j w_j \ . \tag{8.6.25}$$

The squared norm of v is

$$v^2 = v_0^2 + 2\sum_{i=1}^{s} b_i < v_0, w_i > + \sum_{i=1}^{s}\sum_{j=1}^{s} b_i b_j <w_i, w_j> \ . \tag{8.6.26}$$

Taking the inner product of (8.6.24) and w_i yields

$$<v_0, w_i> = <v_i, w_i> - \sum_{j=1}^{s} a_{ij} <w_i, w_j> \ . \tag{8.6.27}$$

The adoption of (8.6.27) in (8.6.26) yields

$$v^2 = v_0^2 + 2\sum_{i=1}^{s} b_i <v_i, w_i> - \sum_{i,j=1}^{s} m_{ij} <w_i, w_j> \ . \tag{8.6.28}$$

If combining

$$<v_i, w_i> = h <Y_i - Z_i \ , f (x_n + hc_i, Y_i)$$

$$\tag{8.6.29}$$

$$- f (x_n + hc_i, Z_i) > \le 0$$

with hypotheses (i) and (ii) of the theorem implies

$$v^2 \le v_0^2 \ ,$$

then the method (8.6.1) is *BN*-stable.   ∎

Definition 8.20. The method (8.6.1) is said to be algebraically stable if

(i) $b_i \ge 0$, $i = 1(1)s$, and

(ii)   the quadratic form $\xi^T M \xi$ is non-negative.

Lemma 8.1. *If ξ is such that $\det(I - A\xi) \ne 0$ and $u = (I - A\xi)^{-1} e$, then*

$$|K(\xi)|^2 - 1 = 1 - 2\sum_{i=1}^{s} b_i \mathrm{Re}(\xi_i)|u_i|^2 - \sum_{i,j=1}^{s} m_{ij}\bar{\xi}_j\bar{u}_i\xi_j u_j \ . \tag{8.6.30}$$

The Lemma 8.1 facilitates the proof of the following theorem:

Theorem 8.4. *The Method (8.6.1) that is algebraically stable is AN-stable. If the c_i, $i = 1(1)s$ are distinct, an AN-stable scheme is algebraically stable.*

Corollary 8.1. *If the coefficients c_i, $i = 1(1)s$ in the Method (8.6.1) are distinct, then AN-stability <=> BN-stability.*

The exciting aspect of this corollary is that it relates a stability property for nonlinear problems directly to a stability property for linear problems.

Butcher (1975) showed that a number of high order classes of Method (8.6.1) are *B*-stable, and he examined the algebraic stability properties of such schemes.

Considering

$$\overline{M} = (\overline{m}_{ij}) ,\qquad\qquad (8.6.31)$$

it can be observed that $A^T \overline{M} A = M$, which is a congruence relation, provided that A is nonsingular.

Table 8.3.
Stability Polynomials for Some Popular Integration Schemes

Method	Stability Polynomial $\Pi(r,z)$
Adams Moulton formula (AMF)	$r^{k-1}(r-1) - z\,\sigma(r)$
Backward Differentiation Formula (BDF)	$\rho(r) - zr^k$
Second Derivative Formula (SDF)	$r^{k-1}(r-1) - z\sigma(r) - \gamma_k z^2 r^k$
Explicit Runge Kutta Formula (RKF) or Polynomial Extrapolation Method	$\sum_{j=0}^{\mu} \alpha_j z^j$
Implicit/Semi-implicit RKF or Rational Extrapolation Method	$\dfrac{\sum_{j=0}^{\mu} \alpha_j z^j}{\sum_{j=0}^{\nu} \beta_j z^j}$

Table 8.4. Test Equations for Stability Criteria

Test Equation	Stability Criterion
$y\prime = \lambda y$	A - and L -
$y\prime = \lambda(x)y$	AN - and LN -
$y\prime = f(y(x))$	B -
$y\prime = f(x,y(x))$	BN- and B -
$y\prime = \lambda(y - g(x)) + g\prime(x)$	S -
$y\prime = A(x)y$	D -

Theorem 8.5. *If A is nonsingular, then Method (8.6.1) is algebraically stable if and only if $b_i \geq 0$, $i = 1(1)s$ and $\overline{M}$ is non-negative definite.*

Butcher (1975) indicated that Method (8.6.1) can attain the order $p = 2s$. Subsequently, a number of authors produced classes of Method (8.6.1) that are algebraically stable and, hence, suitable for stiff systems. Of particular interest are those based on

(a) Gauss-Legendre quadrature of order $p = 2s$ (Butcher 1964),

(b) Radau quadrature IA, IIA of order $p = 2s - 1$ (Ehle 1968), and

(c) Lobatto quadrature IIIC of order $p = 2s - 2$ (Ehle 1969), Chipman 1971).

The implementation difficulties of the implicit Runge-Kutta scheme (8.6.1) have precluded their adoption as a general-purpose algorithm, because every integration step demands the solution of a nonlinear algebraic system of order ms for the Y_i. The computational effort is considerably reduced by adopting method (8.6.1) with $A = (a_{ij})$, a triangular matrix with equal diagonal elements. These diagonally-implicit methods are discussed by Alexander (1977) and extensively studied by Norsett (1974b). Their maximum attainable order is $p \leq s+1$ (cf. Norsett and Wolfbrandt 1979).

The method

$$
\begin{bmatrix}
\lambda & 0 & \lambda \\
1-2\lambda & \lambda & 1-\lambda \\
\dfrac{1}{2} & \dfrac{1}{2} & 0
\end{bmatrix}
\tag{8.6.32}
$$

is algebraically stable *iff* $\lambda \geq 1/4$ with

$$
M = (\lambda - \frac{1}{4})\begin{bmatrix} 1 & -1 \\ -1 & 1 \end{bmatrix} .
\tag{8.6.33}
$$

Furthermore, Method (8.6.32) is algebraically stable with order $p = 3$ if $\lambda = (3 + \sqrt{3})/6$.

Norsett (1974b) constructed a family of three-stage schemes of order $p = 4$:

$$
\begin{bmatrix}
\lambda & & & \lambda \\
\dfrac{1}{2} - \lambda & \lambda & & \dfrac{1}{2} \\
2\lambda & 1-4\lambda & \lambda & 1-\lambda \\
\dfrac{1}{6}(1-2\lambda)^2 & 1-\dfrac{1}{3}(1-2\lambda)^2 & \dfrac{1}{6}(1-2\lambda)^2 & 0
\end{bmatrix} .
\tag{8.6.34}
$$

The s-stage diagonally implicit

$$
\begin{bmatrix}
\lambda & & & & \lambda \\
b_1 & \lambda & & & b_1+\lambda \\
\vdots & & & & \vdots \\
b_1 & b_2 & b_{s-1} & \lambda & \sum_{i=1}^{s-1} b_i+\lambda \\
b_1 & b_2 & & b_s & 0
\end{bmatrix}
\tag{8.6.35}
$$

scheme is algebraically stable $iff\ 0 \le b_i \le 2\lambda$, $i = 1(1)s$, since

$$
M = \mathrm{diag}(b_1(2\lambda - b_1), ..., b_s(2\lambda - b_s)). \tag{8.6.36}
$$

Chipman (1973) and Bickart (1977) examined the problem of efficient implementation of IRK methods. Butcher (1979) suggested the use of a similarity transformation $T^{-1}AT = B$ in which B possesses a much simpler structure that renders the implementation more efficient. The application of this procedure to the method of Burrage (1978a) yielded an IRK of order $p = s$ or $p = s+1$, but B is now a simple Jordan block. Varah (1979) extended the application of this technique to some general IRK that makes the IRK more effective and competitive with the BDF. This he achieved by performing a similarity transformation to the Hessenberg form of the Jacobian matrix, rather than using the LU factorization.

The adoption of the singly implicit R-K scheme whose coefficient matrix has a one-point spectrum reduces the main computational effort to approximately the level of the LMM. It may, therefore, prove to be competitive with the BDF family of codes.

Chapter 9 discusses the recently developed automatic codes for stiff problems which include

(i) Runge-Kutta codes DIRK, SIRK, SIMPLE (cf. Norsett and Thomsen 1984, 1986a,b), based on two B-stable/AN-stable SDIRK methods of order three. Cash (1983c) developed the code MEBDF, based on Modified Extended Backward Differentiation Formula, discussed in Cash (1981a).

(ii) The backward differentiation formulas (BDF, DEPAC, LSODE and its variants):

(iii) the second derivative formulas SECDER, and

(iv) the extrapolation processes IMPEX2, IMPEX3, METAN1, METAN2.

The next section deals with the properties of one-leg multistep methods.

8.7. One-Leg Multistep Methods

Dahlquist (1975) introduced the one-leg multistep method

$$
\rho(E)y_n = hf(\sigma(E)x_n\ ,\ \sigma(E)y_n)\ ,\ \sigma(1) = 1 \tag{8.7.1}
$$

for the numerical solution of the IVP (8.3.13). He observed that the stability analysis of the LMM

$$h^{-1}\rho(E)y_n = \sigma(E)f(x_n, y_n) \tag{8.7.2}$$

can be derived from those of (8.7.1) if constant stepsize h was adopted. This is due to the equivalence property that if $\{y_n\}$ satisfies the one-leg difference equation (8.7.1), then the sequence

$$\{\hat{y}_n \mid \hat{y}_n = \sigma y_n\} \tag{8.7.3}$$

satisfies the LMM

$$\rho(E)\hat{y}_n = h\sigma(E)f(x_n, \hat{y}_n) . \tag{8.7.4}$$

Nevanlinna and Liniger (1978a,b and 1979) further established that the stability regions and the contractivity regions, respectively, for (8.7.2) and (8.7.1) are identical for the linear problem

$$y\prime = \lambda y \tag{8.7.5}$$

or its matrix equivalent

$$y\prime = Ay . \tag{8.7.6}$$

The one-leg implementation is found to be advantageous in terms of storage, since only k values of y need be stored to advance to the next integration step.

Gourlay (1970) and Nevanlinna and Liniger (1978, 1979) used the test problem

$$y\prime = \lambda(x)y , \ \text{Re}\lambda(x) < 0 \tag{8.7.7}$$

to illustrate the superior stability properties of the variable step one-leg methods over the equivalent linear multistep formulas.

While the numerical solution to (8.7.7) generated with the trapezoidal scheme (3.4.1) may be unbounded, its one-leg twin

$$y_{n+1} = y_n + hf(x_{n+1/2}, \ \frac{y_n + y_{n+1}}{2}) \tag{8.7.8}$$

yields stable solutions.

Another significant advantage of the one-leg multistep method is the ease of its application to the differential algebraic equation (DAE)

$$F(x, y, y\prime) = 0 , \tag{8.7.9}$$

which yields the following difference equation

$$F(\sigma x_n, h_n^{-1}\rho(E)y_n, \sigma(E)y_n) = 0 . \tag{8.7.10}$$

Dahlquist (1979a,b) used the autonomous differential system

$$y' = f(y) \, , \, x > a \tag{8.7.11}$$

to establish that the order conditions for the one-leg multistep methods (8.7.1) are, in general, more stringent than those of their linear multistep twin (8.7.2).

The non-autonomous differential system

$$w' = g(x, w) \, , \, w \in R^{m-1} \tag{8.7.12}$$

can be transformed to the autonomous system (8.7.11) by setting

$$y = (x, w(x)^T)^T \tag{8.7.13}$$

and

$$f(y(x)) = (1, g(x, w(x))^T)^T \, . \tag{8.7.14}$$

Theorem 8.13. *A one-leg multistep method (8.7.1) has error order*

$$p = \min\{q, s) \tag{8.7.15}$$

provided

$$\sum_{j=0}^{k} \alpha_j \gamma_j^r = r\bar{h} \sum_{j=0}^{k} \beta_j \gamma_j^{r-1} \, , \, r = 0(1)q \, , \tag{8.7.16a}$$

$$\sum_{j=0}^{k} \beta_j \gamma_j^{r-1} = [\sum_{j=0}^{k} \beta_j \gamma_j]^{r-1} \, , \, r = 1(1)s \, , \tag{8.7.16b}$$

where

$$\gamma_j = x_{n+j} - x_n^* \tag{8.7.16c}$$

and x_n^* is a suitable reference point in the interval $x_n \leq x \leq x_{n+k}$.

For variable step methods, α_j, β_j, $j = 0(1)k$ depend only on the step ratios

$$h_{n+i}/h_{n+i-1} \, , \, i = 2(1)k \, ,$$

and $\bar{h} = \bar{h}_n$ is homogeneous of first degree in h_{n+i}, $i = 1(1)k$. As for the constant step implementation of (8.7.1), $\bar{h} = h$.

The LMM (8.7.2) is insulated from the error order constraint (8.7.16b) imposed on the one-leg multistep twin (8.7.1).

9. STIFF ALGORITHMS

This chapter examines the existing stiff codes based on the implicit/semi-implicit, singly/diagonally implicit Runge-Kutta schemes, the backward differentiation formulas, the second derivative formulas, and the extrapolation processes. Recent attempts to make the other methods robust, like the BDF, are discussed.

9.1. What are Stiff Algorithms?

The best algorithms for stiff initial value problems (in terms of efficiency and reliability) include

(i)   the backward differentiation formulas, BDF (Gear 1969, 1971a,b, 1980c; Hindmarsh 1974, 1980; Byrne and Hindmarsh 1975; Shampine and Watts 1979a; and Petzold 1983) and MEBDF developed by Cash (1983c),

(ii) the second derivative methods (Enright 1972, 1974a,b, 1975; Sacks-Davis 1977, 1980; Sacks-Davis and Shampine 1981; and Addison 1979),

(iii) extrapolation processes (Lindberg 1972, 1973; Bader and Deuflhard 1983; Deuflhard 1983, 1985; and Shampine 1983b-d), and

(iv) the implicit, semi-implicit and singly implicit Runge-Kutta methods (Butcher 1964, 1976a,b; Rosenbrock 1963; Norsett 1974b; Burrage 1978a,b; Burrage, Butcher and Chipman 1980; Wolfbrandt 1977b; Verner 1979; Varah 1979; Cash 1975, 1976, 1977c,d; Cash and Singhal 1982; Bokhoven 1980; Alexander 1977; Kaps and Wanner 1981; Kaps and Rentrop 1979, 1984; Kaps, Poon and Bui 1985; Kaps and Ostermann 1985a,b; Norsett and Thomsen 1984, 1986a,b; Verwer 1981, 1982), and

(v) cyclic or composite multistep methods (Tendler et al. 1978; Bickart et al. 1971; Bickart and Picel 1973; Albrecht 1978, 1985, 1986, 1987; Tischer and Sacks-Davis 1983; Tischer and Gupta 1983a,b, 1985; Iserles 1984; Zhou 1986; Rosser 1967, and Chu and Hamilton 1987).

Enright et al. (1975) suggested that the performance of a stiff method can be assessed with a battery of 25 test problems, which are classified into five distinct categories:

(a) linear with real eigenvalues,

(b) linear with complex eigenvalues,

(c) nonlinear coupling from transient components to smooth components,

(d) nonlinear with real eigenvalues, and

(e) nonlinear with complex eigenvalues.

Enright and Pryce (1983, 1987) included five problems on chemical kinetics as class (f).

Enright et al. (1975) suggested the following input vector for each method:

$$(f,\ a,\ y_0,\ b,\ \tau(x),\ h_{max},\ \frac{\partial f}{\partial y},\ h_0)\ , \qquad (9.1.1)$$

whose first four components describe the initial value problem. $\tau(x)$ is an error control parameter given by

$$\tau(x) = q(x)^*\varepsilon\ , \qquad (9.1.2)$$

where $q(x)$ is the magnitude of the dominant eigenvalue (cf. Lindberg 1974), and ε is the allowable tolerance. The numerical solution is considered acceptable if the errors in the solution satisfy

$$y(x_{n+1}) - y_{n+1\infty} \le (x_{n+1}-x_n)\ \tau(x_{n+1})\ . \qquad (9.1.3)$$

The maximum allowable meshsize h_{max} is included in the input vector so as to ensure that no interesting part of the solution is skipped, while the initial meshsize h_0 is chosen to guarantee that all the interesting transients are followed. Enright et al. (1975) suggested the adoption of

$$h_0 = \frac{1}{|\lambda_{max}|}\ , \qquad (9.1.4)$$

while other possible initial stepsizes are discussed in the latter part of this section.

In a truly automatic code, the left hand side of Equation (9.1.3) should be replaced by an estimate of the local truncation error, although Gear (1980) suggested the replacement of the input vector (9.1.1) by

$$(a,\ y_0,\ b,\ f,\ h_{prin},\ \varepsilon)\ . \qquad (9.1.5)$$

The factors which influence the performance of a numerical integration algorithm include

(i)   the dimension of the differential system,

(ii)   the allowable error tolerance ε,

(iii) the distribution of the eigenvalues of the Jacobian $\partial f / \partial y$, and

(iv)   the coupling between the smooth and the transient components in the system.

The following conclusions were drawn from the performance of the four methods on the battery of 25 test problems:

(i)   The BDF is less efficient than the second derivative methods when the eigenvalues are close to the imaginary axis. However, the implementation of tested Enright's second derivative formulas performs a lot more linear algebra per integration step than the BDF (this cost has been considerably reduced by Enright 1978, and Enright and Kamel 1979).

(ii)   The second derivative methods are less efficient for highly nonlinear problems and nonstiff problems.

(iii) Extrapolation processes (TRAPEX, Enright et al. 1975, and IMPEX2, IMPEX3, Lindberg 1971a,b, 1972, 1973) are less efficient if there exists strong nonlinear coupling between the smooth and transient components, and if stringent error tolerances are imposed. In addition, there is the problem of interpolation at printing intervals, since extrapolation, in general, accommodates the use of larger meshsizes. Enright and Hull (1976) indicate the efficiency of IMPEX2 (Lindberg 1973) at large tolerance. Bader and Deuflhard (1983) have since proposed a superior extrapolation scheme for stiff IVPs in the code METAN1. IMPEX2 does global, passive extrapolation, while METAN1 does local, active extrapolation on a different, much more stable formula.

(iv)   The implicit Runge-Kutta process is considered to be less efficient and less reliable. The situation has since improved considerably with the introduction of

(a)   the semi-implicit Runge-Kutta methods (Norsett 1974b; Cooper and Sayfy 1979; and Norsett and Thomsen 1986a,b),

(b)   the Rosenbrock methods (Rosenbrock 1963; Wolfbrandt 1979; Kaps and Rentrop 1979; Kaps and Wanner 1981; Shampine 1982; and Verner 1979,

(c)   the singly implicit Runge-Kutta methods (Burrage 1978; Butcher 1976; and Burrage et al. 1980),

(d)   the diagonal IRK (Alexander 1977),

(e)   least squares multistep formulas (Gupta 1976, 1985), and

(f)   mono-implicit Runge-Kutta Formulas (Cash and Singhal 1982).

The five basic components of a numerical integration code include

(a)   the identification of the scale of the problem;

(b)   efficient stepsize control, choice of method, and appropriate order;

(c)   estimate of the solution;

(d)   reliable estimate of the local truncation or global error; and

(e)   acceptance or rejection of the solution and error indicator.

A measure of the total cost may be based on the following parameters:

(a)   the CPU time,

(b)   the overhead time of the methods,

(c)   the number of function evaluation,

(d)   the number of steps required,

(e)   the number of Jacobian evaluations, and

(f)   the number of matrix inversions or LU decomposition.

Although the assertions regarding the performance of the stiff methods are based on differential systems of small dimensions, they still hold substantially for large systems. This is due largely to the fact that reliability and the number of function evaluations are independent of the size of the problem. This is not to say, however, that it is always valid to extrapolate conclusions reached from small systems to large systems.

Shampine, Watts, and Davenport (1976) observed that implementation decisions can seriously affect the performance of an integration code.

Possible estimates of the initial stepsize h_0 have been proposed:

(a)   Sedgwick (1973) recommended the use of

$$h_0 = \text{smallest permissible machine unit} . \qquad (9.1.6)$$

(b)   Enright et al. (1975) suggested the adoption of

$$h_0 = \min_j \frac{1}{|\lambda_j|} , \quad |\lambda_j| \neq 0 , \qquad (9.1.7)$$

where $\lambda_j = 1(1)m$ are the eigenvalues of the Jacobian $(\partial f_i/\partial y_j)$ of (2.1.4). Equation (9.1.7) is particularly reasonable if the IVP is stiff.

(c)   Shampine (1977) argued that h_0 be chosen as

$$h_0 = R \left[\frac{\varepsilon^2}{\dfrac{1}{\max(1, |x_0|)^2} + \dfrac{y_0'}{\omega}^2} \right]^{1/2(p+1)} , \qquad (9.1.8)$$

whose parameters are specified as follows:

p = order of method,

ε is the expected accuracy,

$x_0 = a$,

R is a safety factor; $R < 1$, and

$$\frac{y\prime}{\omega} = (\frac{y_1'}{\omega_1}, \frac{y_2'}{\omega_2}, ..., \frac{y_m'}{\omega_m})^T , \tag{9.1.9}$$

where the components of ω are weights used in determining the application of ε to each component of y in the L_2-norm.

(d)   Gear (1980a) suggested the adoption of the initial meshsize

$$h_0 = \min[\varepsilon, \frac{k\varepsilon u^{p/\beta_1}}{\dfrac{y\prime_p}{\omega}}]^{1/2p} \tag{9.1.10}$$

for the fourth-order Runge-Kutta scheme. The parameter u is the relative precision in the calculation of $y\prime$, and k is a method-related constant.

(e)   Norsett and Thomsen (1986b) adopted an initial meshsize

$$h_0 = \frac{(\dfrac{\varepsilon}{c})^{(1/p+1)}}{\sqrt{y_0\prime}} . \tag{9.1.11}$$

The possible risks in the wrong choice of initial meshsize h_0 are that

(i)   if it is excessively small, round-off errors may significantly corrupt the solution, and

(ii)   on the other hand, if it is too large, some local properties of the solution may be omitted. However, too large an initial meshsize should be detected by the step selection mechanism, resulting in a proper step.

Efficient computer implementation of the Adams formulas for nonstiff IVPs have been developed by Krogh (1970), Sedgwick (1973), and Shampine and Gordon (1975). Of historical importance is also Gear's DIFSUB (1971a,b) which implements the Adams-Moulton formulas for nonstiff problems and the backward differentiation formulas (BDF) for stiff systems. This has since been exhaustively revised by Hindmarsh (1979) in the code LSODE and its variants LSODI, LSODA, LSODAR, and LSODES. Shampine and Watts (1979a) also developed an excellent suite of codes that go by the name of DEPAC. It is comprised of the AMF, BDF, and RKF45. Recently, Gear (1980c) developed a computer implementation of the BDF and AMF based on the modified divided difference approach.

Enright and Pryce (1983, 1987) extended both the stiff and nonstiff DETEST problems (Enright et al. 1975 and Hull et al. 1972). A class of five IVPs containing discontinuities was added to the nonstiff set, while a class of five problems on chemical kinetics was added to the stiff DETEST problems.

Enright and Pryce developed testing tools for assessing the efficiency and reliability of a standard numerical method without significantly modifying the method, and without affecting the performance of the method.

9.2. Efficient Implementation of Implicit Runge-Kutta Methods (IRK)

Most of the excellent codes for efficient and reliable solution of initial value problems are the linear multistep formulas based on the Adams-Moulton methods (5.3.2) for nonstiff problems (zones), and the backward differentiation formulas (5.3.18) for stiff problems (zones).

With its superior stability properties and high degree of accuracy, the IRK should constitute an attractive alternative. The major difficulties with the Runge-Kutta processes are that

(a) The explicit R-K processes that can cope with nonstiff problems call for a large number of stages if there is demand for high accuracy. This, in turn, asks for many derivative evaluations—at least s per integration step. In the nonstiff case, explicit R-K methods are nearly competitive with Adams methods, as it is possible to take larger stepsizes with higher order R-K methods. Thus, we might expect that IRK methods could use larger stepsize than the BDF, especially considering the relatively large sizes of the error constants of the BDF. However, the cost of failure is greater for IRK, since it does more function evaluations and perform more linear algebra. Sommeijer (1986) constructed schemes which require fewer derivative evaluations by using some of the past internal approximations.

(b) There is also a general absence of families of R-K methods that can be combined into a variable order sequence with a local error estimator, which, of course, abounds in the linear multistep methods. Although the Richardson extrapolation scheme (3.2.13), which is usually adopted to estimate the local error, is fairly accurate for step-control purposes, it involves additional computational effort of around 50%. A considerable reduction in the computational effort is possible if one adopts either

(i) The approach of Bui and Poon (1981) and Kaps, Poon, and Bui (1985), whereby two Runge-Kutta methods of the same order produce the same Y-vectors, or

(ii) the imbedding methods, whereby an s-stage R-K scheme of order p generates the numerical solution y_{n+1}, while an $(s+1)$-stage scheme of order $p+1$ gives the local error estimate as the difference between the two solutions. Fehlberg (1964, 1968, 1969, 1970) and Verner (1978, 1979) proposed error estimates and stepsize control for the explicit R-K processes, while Burrage (1978) implemented the same idea for implicit R-K schemes.

(c) The s-stage implicit R-K scheme (8.6.1) applied to the IVP

$$y' = f(x, y(x)), \ y(a) = y_0, \ f : R \times R^m \to R^m \tag{9.2.1}$$

results in the s internal approximations of Y_i

$$Y_i = y_n + h\sum_{j=1}^{s} a_{ij} f(x_n + c_j h, Y_j) , \quad i = 1(1)s , \qquad (9.2.2a)$$

and the estimate of the solution

$$y_{n+1} = y_n + h\sum_{j=1}^{s} b_j f(x_n + c_j h, Y_j) . \qquad (9.2.2b)$$

Equation (9.2.2a) constitutes a set of s nonlinear equations in Y_i, each of which is an m-vector. A single application of a variant of the Newton-Raphson scheme to (9.2.2a) yields a modified approximation

$$Y_i^{(r+1)} = Y_i^{(r)} + \delta_i^{(r)} , \quad i = 1(1)s . \qquad (9.2.3)$$

The correction terms $\delta_i^{(r)}$ are given by the linear system

$$B\delta^{(r)} = C^{(r)} , \qquad (9.2.4a)$$

where $ms \times ms$ matrix B

$$B = (B_{ij}) , \quad i, j = 1(1)s$$

has elements

$$B_{ij} = \begin{cases} I - h a_{ij} J_j , & i = j , \\ -h a_{ij} J_j , & i \neq j . \end{cases} \qquad (9.2.4b)$$

C is the matrix

$$C = (C_1^T, C_2^T, ..., C_s^T)^T ,$$

with each

$$C_i = Y_i - y_n - h\sum_{j=1}^{s} a_{ij} f(x_n + c_j h, Y_j) \qquad (9.2.4c)$$

an m-vector, and

$$\delta_i^{(r)} = (\delta_{i1}^{(r)}, \delta_{i2}, ..., \delta_{im}^{(r)})^T ,$$

and

$$J_i = \frac{\delta f}{\delta y}(x_n + c_i h, Y_i^{(0)}) , \quad i = 1(1)s . \qquad (9.2.4d)$$

Even if each Jacobian J_i is approximated by the single matrix J, which is fixed for as long as an acceptable convergence rate is achieved, the general IRK process still requires the solution of a linear system (9.2.2a) of order $ms \times ms$ for each Newton iteration at every integration step. This requires an operation cost of $1/3 (ms)^3$ for LU factorization of B, and $(ms)^2$ arithmetic operations for backward substitution. This is approximately s^3 times as expensive as the equivalent backward differentiation formula, which requires only the LU factorization of the $m \times m$ matrix $I - h\beta J$, plus backward substitution costs.

Norsett (1974b) suggested the use of the semi-implicit Runge-Kutta schemes, whereby matrix A is a lower triangular matrix. In particular, he ensured that A has only one real s-fold eigenvalue λ. The linear system (9.2.4) thus reduces to

$$\begin{bmatrix} I - h\lambda J & 0 & & 0 \\ -ha_{21}J & I - h\lambda J & 0 & 0 \\ \vdots & & & \\ -ha_{s1}J & -ha_{s2}J & \dots & I - h\lambda J \end{bmatrix} \begin{bmatrix} \delta_1^{(r)} \\ \vdots \\ \vdots \\ -\delta_s^{(r)} \end{bmatrix} = C , \qquad (9.2.5)$$

which requires only a single LU factorization of $I - h\lambda J$ and a forward substitution to obtain $\delta_1^{(r)}, \dots, \delta_s^{(r)}$. This approach is, nonetheless, bedevilled by the following problems:

(i)   Norsett (1974b) showed that the resultant semi-implicit scheme has a limited order $p \le s+1$, and, in fact, the highest attainable order is $p = s$ if $s = 2\mu$, $\mu = 2(1)10$.

(ii) It is difficult to construct semi-implicit R-K schemes with large stage number s and high order $p = s$ or $p = s+1$. This is because of the modified order conditions that have to be solved. Automatic codes based on semi-implicit R-K schemes are, hence, almost impossible.

Shampine (1982) developed a code that selects automatically, at every step, an explicit R-K scheme based on RKF45 in a nonstiff step, and a Rosenbrock formula at a stiff step. The explicit R-K scheme is based on the Fehlberg (1970) (4,5) pair (Shampine and Watts 1971). The Rosenbrock scheme is also a semi-implicit scheme of a special nature. Hairer et al. (1982) and Wolfbrandt and Steihaug (1979) described the general semi-implicit scheme as

$$(I - \gamma_i hA)Y_i = hf (x_n+c_ih, \mu_i) + hA\sum_{j=1}^{i-1}\gamma_{ij}Y_j ,$$

$$\mu_i = y_n + \sum_{j=1}^{i-1}a_{ij}Y_j , \qquad (9.2.6)$$

$$y_{n+1} = y_n + \sum_{i=1}^{s}b_iY_i .$$

The parameters γ_i, γ_{ij}, c_i, a_{ij}, and b_i determine the method, and A is an arbitrary matrix with dimension m. The Rosenbrock method is essentially (9.2.6) with

$$A = \frac{\partial f}{\partial y}(x_n, y_n) , \qquad (9.2.7)$$

and each stage Y_i is obtained by solving a linear system with the same matrix. The Rosenbrock method is known to have better stability properties at the expense of having to evaluate the partial derivative at every integration step, whereas the BDF computes the Jacobian (9.2.7) as infrequently as possible. Bui and Poon (1980) and Kaps and Rentrop (1979) derived some Rosenbrock formulas with internal error estimators, a desirable step in making the method practical and competitive with the BDF. Kaps and Rentrop (1979) developed two codes GRK4A and GRK4T, both of order four. Addison (1979) confirmed the efficiency and reliability of both codes on stiff oscillatory problems, as well as on small stiff problems.

Burrage (1978) considered a more general class of implicit Runge-Kutta methods in which A is no longer a triangular matrix, although its characteristic polynomial has a single s-fold zero. This approach eliminates some of the drawbacks which beset the semi-implicit R-K scheme of Norsett (1974b), in that order $p = s+1$ is easily reachable for $s \leq 5$. The construction of variable order, variable step, implicit R-K schemes is, hence, facilitated. Butcher (1976) showed that the new implicit schemes can be as efficiently implemented as semi-implicit ones.

The following assumptions are proposed by Butcher (1964) to simplify the construction of semi-implicit R-K methods:

$$C(p): \sum_{j=1}^{s} a_{ij} C_j^{k-1} = \frac{C_i^k}{k} , \ i = 1(1)s \text{ and } k \leq p ,$$

$$D(p): \sum_{i=1}^{s} b_i C_i^{k-1} a_{ij} = \frac{b_j(1-c_j^k)}{k} , \ j = 1(1)s , \ k \leq p , \qquad (9.2.8)$$

$$B(p): \sum_{i=1}^{s} b_i C_i^{k-1} = \frac{1}{k} , \text{ for all } k \leq p .$$

Butcher (1964) established the following result for an s-stage R-K scheme:

Lemma 8.1. $C(\eta), D(\xi), B(p)$ with $p \leq \xi + \eta + 1, p \leq 2\eta + 2$ *imply the method is of order p.*

Consequently, for $t = 0(1)s$, $C(s-1), D(t), B(s+t) \Rightarrow$ R-K scheme has order $p = s+t$. If it is also true that if the zeros $C_i, i = 1(1)s$ are distinct, then matrix A is similar to some special type of matrix.

Definition 9.1 (Burrage 1978). A singly implicit Runge-Kutta method (SIRK) is a transformed method of order $p \geq s$ whose matrix A has one real s-fold eigenvalue.

Burrage further established the following theorem:

Theorem 9.1. *The family of singly implicit methods of order $p \geq s$ is given by*

$$
\begin{bmatrix}
 & & & c_1 \\
 & & & \cdot \\
V_s & A_s & V_s^1 & \cdot \\
 & & & \cdot \\
 & & & \cdot \\
b_1 & \cdot \quad \cdot \quad \cdot & b_s & c_s
\end{bmatrix} , \tag{9.2.9}
$$

where

$$
(b_1,\dots,b_s) = (1,\dots,\frac{1}{s})V_s^{-1} . \tag{9.2.10}
$$

V_s *is the Vandermonde matrix*

$$
V_s = \begin{bmatrix} 1 & \dots & c_1^{s-1} \\ 1 & & \\ \vdots & & \vdots \\ 1 & \dots & c_s^{s-1} \end{bmatrix} , \quad
A_s = \begin{bmatrix} 0 & 0 & \dots & 0 & \alpha_{1s} \\ 1 & & & & \\ 0 & \dfrac{1}{2} & & & \vdots \\ \vdots & \vdots & & & \\ 0 & 0 & \dots & \dfrac{1}{s-1} & \alpha_{ss} \end{bmatrix} , \tag{9.2.11}
$$

$$
\alpha_{ks} = \frac{(-1)^{s-k}(k^s-1)(s-1)!\lambda^{s-k+1}}{(k-1)!} , \quad k = 1(1)s , \tag{9.2.12}
$$

$$
\lambda \in R - \{0\} ,
$$

with maximum order $p = s+1$.

Theorem 9.2. *The stability (rational) function corresponding to a singly implicit method of order s is given by*

$$
R(z) = (-1)^s \sum_{k=1}^{s} \frac{\lambda^k L_s^{(s-k)}(\frac{1}{\lambda})z^k}{(1-\lambda z)^s} , \tag{9.2.13}
$$

where L_s is the Laguerre polynomial of degree s. If, in addition, $L_{s+1}^{(1)}(1/\lambda) = 0$, then $R(z)$ is the rational function of a singly implicit method of order $s+1$.

Burrage used the Routh-Hurwitz criterion (Lambert 1973a, p. 80) to obtain the ranges of λ, which ensure A-stable SIRK, displayed in Table 9.1.

Currently, one of the most competitive automatic algorithm for numerical solution of stiff IVPs is the family of SIRK methods of Burrage, Butcher, and Chipman (1980) based on Runge-Kutta processes. Burrage (1978) constructed SIRK formulas of orders up to fifteen that are $A(\alpha)$-stable with $\alpha \geq 83°$.

Table 9.1. Range of λ which ensures A-stability of SIRK

s	λ
1	$[1/2, \infty)$
2	$[1/4, \infty)$
3	$[1/3, 1.06858]$
4	$[0.39434, 1.28057]$
5	$[0.24651, 0.36180] \cup [0.42079, 0.47328]$
6	$[0.28407, 0.54090]$

For large differential systems, the SIRK formulas require roughly s times as much linear algebra per iteration, and r times as many function evaluations as the BDF per integration step.

Burrage, Butcher, and Chipman (1980) implemented a family of SIRK formulas in a code named STRIDE, using a transformation proposed by Butcher (1976). Error estimation was handled by embedding an s-stage p-th order formula in an $(s+1)$-stage $(p+1)$-th order formula. The code STRIDE is $A(\alpha)$-stable, $\alpha \geq 83°$, and requires about the same number of $O(m^3)$ operations per step as the BDF. Gaffney (1984) established that the SIRK formulas are only competitive with the BDF for stiff oscillatory problems and that a code for oscillatory problems seems preferable to a code for stiff (decaying) solutions. Actually, BDFs are unsuitable, and STRIDE is much better for such problems.

Norsett (1974b) and Alexander (1977) proposed the diagonally-implicit R-K (SDIRK) formulas, which belong to a weaker class than the SIRK formulas, with all the non-zero diagonal elements of the coefficient matrix A having a fixed constant λ. This approach demands only the solution of one iteration matrix for the internal functions Y_i.

Shampine (1982) proposed a type-insensitive code called DEGRK based on a pair of A-stable Rosenbrock formulas of orders three and four, with damping at infinity for the stiff steps and the Runge-Kutta-Fehlberg pair in the nonstiff steps.

Gaffney (1984) evaluated the performance of the codes STRIDE, BLENDED, DIFSUB, STINT, DIRK, and LSODE on a linear stiff oscillatory problem—example B5 of Enright et al. (1975):

$$y_1' = -10y_1 + 100y_2 ,$$

$$y_2' = -100y_1 - 10y_2 ,$$

$$y_3' = -4y_3 ,$$

$$(9.2.14)$$

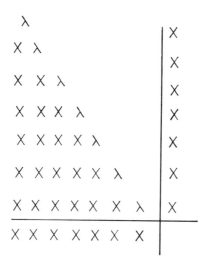

Figure 9.1: Diagonally Implicit R-K Processes (DIRKs)

$$y_4' = -y_4 , \qquad y_i(0) = 1 , \; i = 1(1)6 ,$$

$$y_5' = -0.5y_5 , \qquad \alpha = 100 ,$$

$$y_6' = -0.1y_6 ,$$

whose eigenvalues are -0.1, -0.5, -1, -4, and $-10 \pm 100i$.

Gaffney (1984) made the following observations:

(i)   that the STRIDE and DIRK codes appear to be better suited to stiff oscillatory problems, perhaps because they possess a stronger property of S-stability (see Definition 8.8),

(ii)   that the performance of the DIRK is very impressive on the linear problem,

(iii)   that STRIDE is more efficient on highly nonlinear problems where function evaluations are more expensive, perhaps due to the fact that singly implicit methods can attain higher order $1 \leq p \leq 15$, and

(iv)   that it is quite possible to obtain reasonable efficiency from the code LSODE, using maximum permissible order $p = 5$.

Table 9.2. Methods for Stiff Oscillatory IVPs

Code	Author	Method	Comments
STRIDE	Burrage (1978)	order, based on singly implicit R-K methods of order p in $1 \leq p \leq 15$	Variable step, variable for both stiff and nonstiff options
BLEND DIFSUB	Skeel/Kong (1977)	Variable step, variable order code; it blends the second derivative schemes of Enright (1974) with Adams-Moulton scheme; order $p: \leq p \leq 7$	Can handle both stiff and nonstiff problems
STINT	Tendler et al. (1978)	Variable step, variable order code which uses cyclic composite multistep formulas; order p: $1 \leq p \leq 7$	No non-stiff option
DIRK	Alexander (1977)	Variable step, fixed order code using diagonally implicit R-K formulas; order p: $1 \leq p \leq 4$	Nonstiff option not provided
LSODE	Hindmarsh (1980)	Variable step, variable order code which incorporates Adams-Moulton of order p: $1 \leq p \leq 12$ for nonstiff phase and BDF of order $1 \leq p \leq 5$ for stiff phase	Provision for full and banded Jacobian; provides for both stiff and nonstiff option

Table 9.2 continued

Code	Author	Method	Comments
DSTIFF	Gupta (1985)	Variable step, variable order based on least squares multistep formulas of order $1 \leq p \leq 7$	No provision for sparse or banded Jacobian
SIMPLE	Norsett and Thomsen (1986a,b)	Variable step, variable order code based on SDIRK order $p = 3$,	Provision to handle Stiff and nonstiff
MEBDF	Cash (1983)	Variable step, variable order code based option on extended BDF order $2 \leq p \leq 8$	No non-stiff
ODIOUS	Tischer and Gupta (1983)	Variable step, variable order cyclic option multistep method	No non-stiff

9.3. The Backward Differentiation Formula (BDF)

The BDF can be described as

$$y_{n+k} = h\beta_k f(x_{n+k}, y_{n+k}) + g , \qquad (9.3.1)$$

where g is a constant for the current step, obtained as a linear combination of past values of y (see Section 5.3). As $\partial f / \partial y$ is usually large, stiff codes normally adopt a modified Newton scheme to solve the nonlinear Equation (9.3.1); i.e.,

$$(I - h\beta_k J)(y_{n+k}^{[s+1]} - y_{n+k}^{[s]}) = -y_{n+k}^{[s]} + h\beta_k f(x_{n+k}, y_{n+k}^{[s]}) + g , \qquad (9.3.2)$$

where the matrix $I - h\beta_k J$ is called the Newton iteration matrix, and J is an approximation to the Jacobian.

Most codes recompute the Newton iteration matrix only when there is a problem with convergence or there is a change in stepsize. The linear algebra in (9.3.2) is effected by direct LU decomposition using LINPACK software. Gear and Saad (1983) suggested iterative methods as an alternative. The advantage of this approach is seen in problems with large dimensions and sparse Jacobian matrices, which are of frequent occurrence when methods of lines are used to solve partial differential equations.

Shampine (1979, 1980) examined the problems of solving the nonlinear system (9.3.1), which arises from using implicit methods for stiff initial value problems, while Krogh and Stewart (1983) discussed the effect of using inaccurate Jacobian on the rate of convergence. Carver and MacEwen (1981) proposed sparse matrix techniques for large stiff systems.

Enright (1978) suggested the adoption of the scaled iteration matrix

$$\overline{W}_n = J - (\frac{1}{h_n \beta_k})I \ , \tag{9.3.3}$$

whereby any alteration in h_n or β_k is effected by multiplying the r.h.s. of (9.3.3) by a constant diagonal matrix. Enright further suggested that the scaled iteration matrix be decomposed as

$$\overline{W}_n = LHL^{-1} \ , \tag{9.3.4a}$$

where L is lower triangular matrix, and H is the Hessenberg matrix. If $h_n \beta_k$ is replaced by $\overline{h}_n \overline{\beta}_k$, its effect on (9.3.3) can be expressed as

$$J - (\frac{1}{\overline{h}_n \overline{\beta}_k})I = (J - (\frac{1}{h_n \beta_k})I) + (\frac{1}{h_n \beta_k} - \frac{1}{\overline{h}_n \overline{\beta}_k})I$$

$$\tag{9.3.4b}$$

$$= L[H + (\frac{1}{h_n \beta_k} - \frac{1}{\overline{h}_n \overline{\beta}_k})I]L^{-1} \ .$$

Enright's factorization can achieve considerable savings if the Jacobian is constant or is slowly varying, and if the dimension of the differential system is large in the event of a change in stepsize or order of the method.

Enright and Kamel (1979) went a step further in proposing an iteration scheme based on the LHL^{-1} factorization. They suggested automatically partitioning H into stiff and nonstiff components and then approximating the nonstiff component by either a zero matrix or a diagonal matrix.

Table 9.3. Angle of $A(\alpha)$-Stability of the BDF and MEBDF

order	1	2	3	4	5	6	7	8
BDF	90.0	90.0	86.0	73.4	51.8	17.9	-	-
MEBDF	90.0	90.0	90.0	88.4	83.1	74.5	62.0	43.0

Table 9.4. Automatic Codes for Stiff/Nonstiff Differential Systems

Year	Author	Implementation	Coefficient	Code
1969	Gear	Nordsieck	Fixed	DIFSUB
1972	Brayton et al.	Nordsieck	Variable	
1974	Hindmarsh	Nordsieck	Fixed	GEAR
1975	Sherman/ Hindmarsh	Nordsieck	Fixed	GEARS
1976	Hindmarsh	Nordsieck	Fixed	GEARIB
1976	Hindmarsh/ Byrne	Nordsieck	Variable	EPISODE
1977	Hindmarsh	Nordsieck	Fixed	GEARIB
1980	Hindmarsh	Nordsieck	Fixed	LSODE,LSODI, LSODES
1980*	Shampine/ Watts	Nordsieck	Variable	DEBDF
1980	Curtis	Nordsieck	Fixed	FACSIMILE
1980	Gear	DDF	Variable	
1982	Petzold/ Hindmarsh	Nordsieck	Fixed	LSODA,LSODAR
1983	Cash	BDF	Variable	MEBDF
1985	Gupta	Nordsieck	Variable	DSTIFF

The BDF of orders one and two were first applied to stiff IVPs by Curtis and Hirschfelder (1952). Gear (1969, 1971a,b) implemented the BDF of orders one through six jointly with the AMF of orders one through twelve in a variable order variable stepsize code called DIFSUB. Subsequent versions of DIFSUB are illustrated in Table 9.4. In these versions, the BDF is restricted to orders one through five, since the angle of stability of the sixth order BDF is too small, as is evident from Table 9.3.

The BDF/AMFs are easily the most tested and most developed automatic integrators for stiff/nonstiff problems. Petzold (1983) advocated the adoption of the AMF (using functional iterations) in the transient phase and the BDF (using chord iterations) in the stiff phase. The FORTRAN codes GEAR and LSODE and their variants are greatly improved versions of DIFSUB, and were developed by Hindmarsh (1974, 1980). They both accommodate various iteration schemes.

LSODE combines the capabilities of GEAR and GEARB in that, in case of a stiff differential system

$$y' = f(x, y),$$ (9.3.5)

it treats the Jacobian $J = \partial f / \partial y$ as either full or banded. It is, indeed, flexible and convenient.

The variant LSODI, which supersedes GEARIB, treats systems in the linearly implicit form

$$A(x, y)y' = g(x, y),\qquad(9.3.6)$$

where A is a square matrix that is permitted to be singular, provided that initial values of y and y' are available.

LSODES, which supersedes GEARS, solves (9.3.5), whose Jacobian is a general sparse matrix. The linear system is solved using the Yale Sparse Matrix Package (YSMP), developed by Eisenstat et al. (1977a,b).

While LSODA is an automatic method selection code based on Petzold (1982), the code LSODAR combines its capabilities and, in addition, incorporates a root-finder to identify discontinuities. Hindmarsh (1979, 1983a,b) discussed the various versions of LSODE.

The code GEAR is a direct descendant of DIFSUB; LSODE is descendant of GEAR; DEBDF is a descendant of LSODE. The software development in going from LSODE to DEBDF is as great as that going from GEAR to LSODE.

In order to facilitate an easy change of stepsize and order, either the Nordsieck (1962) history array or the modified divided difference approach (Krogh 1974, Krogh and Stewart 1983, and Gear 1980) can be adopted.

The implementation of the k-step BDF using the Nordsieck vector array requires the storage of the $(k+1)$-vector

$$(y_n, h_n y_n', \frac{1}{2}h_n^2 y_n'', ..., \frac{1}{k!}h_n^k y_n^{(k)})^T .\qquad(9.3.7)$$

Only minor changes in the coefficients are required to switch from BDF to AMF and vice-versa.

Gear (1980c) implemented both the BDF and the AMF, using a unified modified divided difference approach such that only the MDD of y' need be stored. This approach permits easy transition from AMF to BDF, and vice-versa, at a relatively insignificant cost, thus ensuring the competitiveness of the MDD with that of Nordsieck array.

The three basic techniques normally used in the implementation of the variable step linear multistep methods are

(i)   fixed coefficient formulas (Gear 1971a,b and Hindmarsh 1974, 1980),

(ii) variable coefficient formulas (Brayton et al. 1972 and Byrne and Hindmarsh 1975), and

(iii) fixed leading coefficient formulas (Jackson and Sacks-Davis 1980 and Gear 1980).

As for the fixed coefficient approach, interpolation is used to generate approximations to the solution at evenly-spaced past points. The coefficients are normally precomputed, but require interpolation to effect a change in stepsize.

Hence, the computational effort depends on the number of equations in the differential system. This approach is prone to instability if there are frequent stepsize changes, and, hence, codes (typically) keep the stepsize and order unchanged for $p+1$ steps while using a BDF of order p.

In the case of the variable coefficient formulas, the past values are not necessarily evenly-spaced and the coefficients have to be re-calculated if the stepsize has altered during the past k steps. The required computational efforts for the variable coefficient approach depends only on the stepsize and is independent of the number of equations in the system.

Both empirical and theoretical results confirm the superior stability properties of the variable coefficient implementation (Brayton et al. 1972, Gear and Tu 1974, and Gear and Watanabe 1974). Curtis (1980, p. 73) disagreed with this point of view and asserted that it seems more a question of the implementation. Although the variable coefficient approach allows the possibility of stepsize changes at each integration step, it must be noted that each time the stepsize is altered, $h\beta_k$ will change for the next k steps and, hence, result in several re-evaluations of the Newton iteration matrix

$$W_n = I - h_n\beta_k J . \tag{9.3.8}$$

This is where Enright's representation of W_n in (9.3.3) and (9.3.4) comes in handy.

In an attempt to eliminate the need for frequent evaluations of the Newton iteration matrix, Jackson and Sacks-Davis (1980) and Gear (1980c) proposed the fixed leading coefficient approach, in which β_k is kept constant on steps after a change in stepsize. The stability properties of this approach are quite close to those of the variable coefficient approach.

DIFSUB, its variants LSODE and DEBDF, and the unified divided difference scheme (Gear, 1980c) have been thoroughly tested and are known to be robust and reliable. There is abundant promise for LSODA with Petzold's (1983) idea of automatic switch from stiff to nonstiff algorithm and vice versa.

The general trend in the drive toward automation is the development of type-insensitive codes based on the various methods. The codes automatically identify the nature of the problem (either stiff or nonstiff) and opt for the optimum method, order, and stepsize for the next integration step.

9.4. Second Derivative Formulas (SDFs)

Enright (1972, 1974a,b) proposed the second derivative multistep methods

$$y_{n+k} = y_{n+k-1} + h\sum_{j=0}^{k}\beta_j y'_{n+j} + h^2\gamma_k y''_{n+k} \tag{9.4.1}$$

for the autonomous differential system

$$y' = f(y) , y(a) = y_0 . \tag{9.4.2}$$

Most stiff IVPs encountered are independent of time. Besides, those that are not can be readily transformed into autonomous systems by adjoining one extra equation. The choice of the integration formula (9.4.1) is due to the following, desirable properties that such schemes possess:

(i) Excellent stability at the origin and in the neighborhood of $-\infty$. The integration formula is A-stable up to order four and $A(\alpha)$-stable up to order nine.

(ii) High attainable order $p = k+2$, instead of $p = k$, for the BDF, and smaller truncation error coefficients than the BDF.

(iii) Since (9.4.1) is implicit and $h\partial f /\partial y \gg 1$ for stiff IVPs, the integration formula has to be solved with a modified Newton-Raphson iteration to give

$$W_n (y_{n+k}^{[s+1]} - y_{n+k})^{[s]} = - y_{n+k}^{[s]} + h\beta_k f(y_{n+k}^{[s]}) + h^2 \gamma_k \frac{\partial f}{\partial y}(y_{n+k}^{[s]})$$

(9.4.3)

$$+ y_{n+k-1} + h\sum_{j=0}^{k-1} \beta_j y'_{n+j} , \ s = 0, 1, 2,...,$$

with the Newton iteration matrix W_n obtained as

$$W_n = I - h\beta_k J - h^2 \gamma_k J^2 , \tag{9.4.4}$$

where J is some approximation to the Jacobian $(\partial f /\partial y)$ at the current point.

The implementation of Enright's scheme performs more linear algebra per integration step than the linear algebra required by the BDF, as the iteration matrix (9.4.4) requires a matrix multiplication J^2, and the iteration requires a matrix-vector multiplication $Jf(y)$.

However, because of their superior stability properties, the SDFs perform better than the BDF when solving stiff systems whose eigenvalues have large imaginary parts (stiff oscillatory problems) (cf. Enright et al. 1975).

In his code SDBASIC, Enright (1972) adopted a variable step, variable order implementation using a one-step, two-step error estimate, restricting stepsize adjustment to halving, and doubling. Addison (1979) and Sacks-Davis (1980) proposed a more efficient error estimator using the predictor-corrector approach, with error estimates obtained from the difference between the predictor and the corrector.

Enright (1974a) proposed making his code SDBASIC more efficient by eliminating the matrix multiplication J^2, which not only requires $O(m^3)$ operations but can also introduce non-zero coefficients if J is sparse.

By introducing the constraint

$$\beta_k^2 + 4\gamma_k = 0 , \tag{9.4.5}$$

the iterative formula (9.4.3) is transformed into

$$\overline{W}_n(y_{n+k}^{[s+1]} - y_{n+k}^{[s]}) = -y_{n+k}^{[s]} + h\beta_k f(y_{n+k}^{[s]})$$

(9.4.6)

$$-h^2(\frac{\beta_k}{2})^2 Jf(y_{n+k}^{[s]} + y_{n+k-1} + h\sum_{j=0}^{k-1}\beta_j y'_{n+j}\ ,$$

with the iteration matrix specified as

$$\overline{W}_n = (I - \frac{h\beta_k J}{2})^2\ .$$

(9.4.7)

This procedure eliminates the need for matrix multiplication J^2, which not only saves an additional $O(m^3)$ operation, but enables one to take advantage of sparsity of J. One need only perform two back substitutions for each iteration.

The resulting integration formula has a reduced order $p = k+1$ and possesses poorer $A(\alpha)$-stability properties.

Sacks-Davis and Shampine (1981), in a type-insensitive code, showed how the SDF can be used to solve both stiff and nonstiff initial value problems.

9.5. Extrapolation Processes for Stiff Systems

In Chapter 7, various extrapolation schemes for nonstiff systems were discussed. Of particular interest is the Gragg-Bulirsch-Stoer algorithm, reckoned to be among the best codes for such problems. Two extrapolation processes for stiff systems have been widely considered based on

(i)   the trapezoidal rule and the implicit midpoint rule (Lindberg 1971a,b, 1972, 1973) and

(ii) the semi-implicit midpoint rule (Bader and Deuflhard 1983, Deuflhard 1985, Shampine 1983b,c,d; and Shampine 1985a).

Implicit Midpoint Algorithms (Lindberg 1973)

Step 1. Let $x_0 = a, x_N = b, x_n = a+nh, n = 0(1)N$.

Compute $y(x_n + h, h)$ and $y(x_n + h/2, h/2)$ from

$$y(x_n+h,h) = y(x_n,h) + hf(\frac{[y(x_n+h,h) + y(x_n,h)]}{2})\ ,$$

(9.5.1a)

$$n = 0(1)N\ ,$$

and

$$y(x_n+\frac{h}{2},\frac{h}{2}) = y(x_n,\frac{h}{2}) + \frac{h}{2}f(\frac{[y(x_n+\frac{h}{2},\frac{h}{2}) + y(x_n,\frac{h}{2})]}{2})\ ,$$

(9.5.1b)

$$n = 0(1)2N .$$

Step 2. Perform passive smoothing:

$$\hat{y}(x_n, h) = \frac{[y(x_n + h, h) + 2y(x_n, h) + y(x_n - h, h)]}{4} , \tag{9.5.2a}$$

$$\hat{y}(x_n, \frac{h}{2}) = \frac{[y(x_n + \frac{h}{2}, \frac{h}{2}) + 2y(x_n, \frac{h}{2}) + y(x_n - \frac{h}{2}, \frac{h}{2})]}{4} . \tag{9.5.2b}$$

Both (9.5.2a) and (9.5.2b) have asymptotic expansion in h^2.

Step 3. Perform passive extrapolation on the smoothed values

$$\bar{y}(x_n, h) = \hat{y}(x_n, \frac{h}{2}) + \frac{[\hat{y}(x_n, \frac{h}{2}) - \hat{y}(x_n, h)]}{3} . \tag{9.5.3}$$

These values are not used in subsequent computations.

The nonlinear systems of equations are normally solved with the quasi-Newton method:

If $z = y(x_n + h, h)$, $y = y(x_n, h)$, then (9.5.1a) becomes

$$F(z) = z - y - hf(\frac{[z + y]}{2}) = 0 , \tag{9.5.4}$$

which is solved iteratively as

$$W_n(z^{[s+1]} - z^{[s]}) = -F(z^{[s]}) , \tag{9.5.5}$$

where W_n is an approximation to the matrix

$$\frac{\partial F}{\partial z} = I - \frac{h}{2}(\frac{\partial f(\eta)}{\partial \eta})_{\eta = z + y/2} . \tag{9.5.6}$$

The same W_n is usually adopted for several integration steps, until problems of convergence arise or the rate of convergence is rather slow, i.e.,

$$\rho = \frac{z^{[s+1]} - z^{[s]}}{z^{[s]} - z^{[s-1]}} > 0.2 . \tag{9.5.7}$$

The nonlinear Equation (9.5.1b) is solved accordingly.

Semi-implicit Midpoint Rule

Bader and Deuflhard (1983) proposed the semi-implicit midpoint rule combined with polynomial extrapolation for autonomous stiff system

$$y' = f(y) , \ y(a) = y_0 . \tag{9.5.8}$$

The Gragg modified midpoint rule (7.2.11) is known to have an asymptotic h^2 expansion of the form

$$y(x, h_r) - y(x) = \sum_{j=1}^{N} [u_j(x) + (-1)^{N_r} v_j(x)] h^{2j} + E_{N+1}(x; h) h^{2N+2} \qquad (9.5.9)$$

where $N_r = h/h_r$, $r = 0(1)M$, M depends on the differentiability properties of f, and $u_j(x)$, $v_j(x)$ satisfy the linear systems

$$u_j' = f_y(y(x)) u_j + ...,$$

$$\qquad\qquad (9.5.10)$$

$$v_j' = -f_y(y(x)) v_j + ...$$

Suppose (9.5.8) has the following theoretical solution:

$$y(x) = e^{-x} + e^{-2000x} . \qquad (9.5.11)$$

If x is sufficiently large, (9.5.11) reduces to

$$y(x) \equiv e^{-x} .$$

But integrating (9.5.10) yields

$$u_j(x) = e^{-x} , \ v_j(x) = -e^{+2000x} , \qquad (9.5.12)$$

thus depicting the weak stability of the component $v_j(x)$.

Bader and Deuflhard (1983) applied the transformation (in the spirit of Lawson 1966, 1967a,b)

$$y(x) = e^{Ax} c(x) \qquad (9.5.13)$$

to (9.5.8) to get the system

$$c'(x) = e^{-Ax} \bar{f}(y) , \qquad (9.5.14)$$

where

$$\bar{f}(y) = f(y) - Ay . \qquad (9.5.15)$$

They applied the explicit midpoint rule to (9.5.14) and then substitute

$$c(x) = e^{-Ax} y(x)$$

to get

$$\bar{z}_0 = y_n ,$$

$$\bar{z}_1 = e^A h_r(y_n + h_r \bar{f}(y)) ,$$

$$\qquad\qquad (9.5.16)$$

$$\bar{z}_{s+1} = e^A h_r(e^A h_r \bar{z}_{s-1} + 2 h_r \bar{f}(\bar{z}_s)) , \ s = 1(1)N_r ,$$

$$\bar{y}(x_{n+1}, h_r) = \frac{1}{4}[e^{A h_r} - z_{N_r-1} + 2\bar{z}_{N_r} + e^{-A h_r} z_{N_r+1}] .$$

It can be shown that (9.5.16) has asymptotic expansion in h^2; i.e.,

$$y(x_{n+1}, h) - y(x_n) = \sum_{j=1}^{N} \bar{u}_j(x) + (-1)^{N_r} \bar{v}_j(x)]h^{2j} \qquad (9.5.17)$$

$$+ \bar{E}_{N+1}(x; h)h^{2N+2} ,$$

where

$$\bar{u}_j' = f_y(y(x))\bar{u}_j +..., \text{ and} \qquad (9.5.18)$$

$$v_j' = f_y(y(x))\bar{v}_j + 2(A - f_y)\bar{v}_j +.... \qquad (9.5.19)$$

With

$$A = f_y(y_n) , \qquad (9.5.20)$$

the Example (9.5.11) yields

$$\bar{u}_j(x) = e^{-x} , \ v_j'(x) = e^{-x} , \qquad (9.5.21)$$

whereby the previous instability has been eliminated.

With the notation

$$J = f_y(y_n) , \ \bar{f}(y) = f(y) - Jy , \qquad (9.5.22)$$

Bader and Deuflhard (1983) proposed the semi-implicit algorithm to generate the zero-th column of the extrapolation table

$$\eta_0 = y_n ,$$

$$\eta_1 = (I - h_r J)^{-1}[y_n + h_r \bar{f}(y_n)] , \qquad (9.5.23a)$$

$$\eta_{s+1} = (I - h_r J)^{-1}[(I + h_r J)\eta_{s-1} + 2h_r \bar{f}(\eta_s)] , \ s = 1(1)N_r , \qquad (9.5.23b)$$

and

$$y(x_{n+1}, h_r) = \frac{1}{2}(\eta_{N_r-1} + \eta_{N_r+1}) . \qquad (9.5.23c)$$

Bader and Deuflhard (1983) established that, under certain conditions, $y(x_{n+1}, h_r)$ obtained from (6.5.23) has an error expansion in h^2.

Theorem 9.3. *Let y denote the unique solution of the IVP (9.5.8), and $y(x; h_r) = \eta_{N_r}$ the numerical estimate generated by the semi-implicit algorithm (9.5.23) for $x = N_r \cdot h_r$ fixed. Assume that for some inner product $<\cdot>$, the following relation holds:*

$$<x, Jx> \le \mu<x, x> \equiv \mu x^2 . \tag{9.5.24}$$

If $f \varepsilon C^{2N+2}$ and

$$\bar{f}(u) - \bar{f}(v) \le \bar{L}u - v , \tag{9.5.25}$$

then there exists some $H > 0$ subject to

$$H\mu \le 1 - \xi \tag{9.5.26}$$

for some ξ in the range of $0 < \xi < 1$ such that, for all $h \ \varepsilon[0, H]$, the asymptotic error expansion in h^2 holds

$$y(x; h) - y(x) = \sum_{j=1}^{N} [u_j(x) + (-1)^{N_r} v_j(x)]h^{2j} + E_{n+1}(x,h)h^{2N+2} . \tag{9.5.27}$$

The functions u_j, v_j (excluding the inhomogeneous terms) satisfy

$$u'_j = f_y(y)u_j +...,$$

$$\tag{9.5.28}$$

$$v'_j = f_y(y)v_j + 2(A - f_y(y))v_j +...,$$

with E_{n+1} uniformly bounded for all $h \ \varepsilon[0, H]$.

Bader and Deuflhard (1983) showed that the stability function for η_{2m-1} is

$$\eta_{2m-1} = \frac{1}{1-z}[\frac{1+z}{1-z}]^{m-1} , \tag{9.5.29a}$$

and for η_{2m} is

$$\eta_{2m} = [\frac{1+z}{1-z}]^m , \tag{9.5.29b}$$

while that of $y(x; h)$ is

$$\frac{1}{(1-z)^2}[\frac{1+z}{1-z}]^m , \tag{9.5.29c}$$

thus exhibiting the A-stability of the semi-implicit midpoint scheme.

Efficient Implementation of Semi-Implicit Midpoint Rule

An even number N_r of extrapolation steps is normally adopted because of the smoothing procedure (9.5.23c), and a convergence condition requires that

$$\frac{N_r}{N_{r+1}} \le \alpha < 1 . \tag{9.5.30}$$

Bader and Deuflhard suggested the sequence $\{N_r\}$ given by

$$F_\alpha = \{2, 6, 10, 14, 22, 34, 50, 70, 98, 138,...\} \tag{9.5.31}$$

in their code METAN1. This defines the sequence of decreasing meshsizes $\{h_r\}$ with

$$h_r = \frac{h}{N_r} , r = 1, 2,.... \tag{9.5.32}$$

With the identity

$$(I - hJ)^{-1}(I + hJ) = 2(I - hJ)^{-1} - I \tag{9.5.33}$$

and the variables

$$\Delta_s = \eta_{s+1} - \eta_s , \tag{9.5.34}$$

the semi-implicit midpoint rule (9.5.23) can be compactly implemented to generate an estimate of $y(x_{n+1})$ as

$$\eta_0 = y_n ,$$

$$\Delta_0 = (1 - h_r J)^{-1} h_r f (y_n) ,$$

$$\eta_s = \eta_{s-1} + \Delta_{s-1} ,$$

$$\tag{9.5.35}$$

$$\Delta_s = \Delta_{s-1} + 2(I - h_r J)^{-1}[h_r f (\eta_s) - \Delta_{s-1}] , s = 1(1)N_r - 1 ,$$

$$\overline{\Delta}_{N_r} = (I - h_r J)^{-1}[h_r f (\eta_{N_r}) - \Delta_{N_r-1}] ,$$

$$y (x_{n+1}, h_r) = \eta_{N_r} + \overline{\Delta}_{N_r} .$$

Estimate (9.5.35) only requires one Gaussian decomposition of $(I - hJ)^{-1}$, as well as forward substitutions and backward substitutions in every integration interval.

Bader et al. suggested some scaling of f_y, which results in a scaling of the arising linear system as

$$D^{-1}(I - hf_y)D = I - h(D^{-1}f_yD) \tag{9.5.36}$$

for some nonsingular (diagonal) weighted matrix D. A matrix is considered singular if one encounters an exactly zero pivot or

$$D^{-1}\Delta > 1 . \tag{9.5.37}$$

Shampine (1983c) developed a type-insensitive code METAN2, in which $J = 0$ in (9.5.35) for nonstiff zone, thus yielding a variant of the explicit midpoint rule, the only variation being the smoothing pattern. Shampine (1983b,c,d, 1985a) and Shampine, Baca, and Banner (1983) suggested ways to improve the efficiency of extrapolation methods.

9.6. Mono-Implicit R-K Formulas

Cash and Singhal (1982) extended (4.6.20), a general one-step integration formula given as

$$y_{n+1} = y_n + h \sum_{i=1}^{s} b_i f(x_{n+c_i}, y_{n+c_i}), \tag{9.6.1a}$$

$$y_{n+c_i} \approx \overline{y}_{n+c_i} = \delta_i y_{n+1} + (1-\delta_i) y_n + h \sum_{j=1}^{m(i)} a_{ij} f(x_{n+c_j}, y_{n+c_j}), \text{ and} \tag{9.6.1b}$$

$$c_i = \delta_i + \sum_{j=1}^{m(i)} a_{ij}, \, m_i \leq s. \tag{9.6.1c}$$

Equations (9.6.1a-c) can either be viewed as a contracted linear multistep method with stepsize h rather than sh, as a hybrid one-step method, or even as a collocation method.

Table 9.5 illustrates how various modes of the R-K methods are represented by (9.6.1).

Cash (1975b) discussed the computational advantages of the MIRK formulas over the IRK methods, and he proposed some L-stable methods based on (9.6.1). Bokhoven (1980) also proposed A-stable MIRK methods, all of which require significantly more computational efforts than the BDFs. His A-stable method of order six is described as

$$y_{n+1} = H(x_n, y_n, y_{n+1}), \tag{9.6.3}$$

which is solved by Newton's method as

$$(I-B)(y_{n+1}^{(t+1)-y_{n+1}^{(t)}}) = H(x_n, y_n, y_{n+1}^{(t)}) - y_{n+1}^{(t)}, \tag{9.6.4a}$$

$$B = h \sum_{i=1}^{s} b_i V_i, \, V_i = J_i(\delta_i I + h \sum_{j=1}^{m(i)} a_{ij} V_j), \tag{9.6.4b}$$

where $J = f_y$ at the point (x_{n+c_i}, y_{n+c_i}).

In practice, modified Newton iteration is effected with $J_i = J(x_{n+1}, y_{n+1})$, $i = 1(1)m(i)$, with

Table 9.5. Coefficients of Runge-Kutta Formula (9.6.1)

MODE	Constraints of Coefficients	
Explicit	$m(i) = i-1$, $\delta_i = 0$	(9.6.2a)
Diagonally Implicit	$m(i) = i$, $\delta_i = 0$, $a_{ii} = \beta_i$ for all i	(9.6.2b)
Fully Implicit	$m(i) = s$, $\delta_i = 0$	(9.6.2c)
Semi Implicit	$\begin{cases} m(i) = i-1,\ \delta_i = 0,\ 1 \le i \le r < [\tfrac{1}{2}\,s] \\ m(i) = i-1,\ \delta_i = 1,\ a_{ij} = 0, \end{cases}$	
	$r+1 \le i \le s$, $1 \le j \le r$	(9.6.2d)
Mono Implicit	$m(i) = i-1$, $\delta_i \ne 0$ for at least one i	(9.6.2e)

$$B = \frac{1}{2}\, h\, J - \frac{1}{10} h^2 J^2 + \frac{1}{120} h^3 J^3. \qquad (9.6.5)$$

Apart from the fact that the evaluation of B in (9.6.5) requires $2m^3$ multiplications, disadvantages of the MIRK include a possible loss of sparsity in adopting the iteration matrix $(I - B)$ (if even J is sparse). Besides, $I - B$ may be singular due to loss of lower order terms. The meshsize may be reduced to avoid singularity.

Bokhoven (1980) suggested the adoption of the iteration matrix $B = \tfrac{1}{2} hJ$ instead of (9.6.5), but this approach calls for the use of a small meshsize h in order to obtain convergence. Cash (1976) suggested that the coefficients of (9.6.1) be chosen to ensure that

$$I - B = \prod_{i=1}^{s} (I - \alpha h\, J_i)\,, \qquad (9.6.6)$$

but the resultant scheme is still less efficient than a corresponding LMM.

Cash and Singhal (1982) derived some embedded L-stable MIRK formulas of orders up to four, in which the iteration matrix of the modified Newton scheme is replaced by a suitable matrix that factorizes as a power of a single matrix; i.e.,

$$I - B \approx (LU)^t = (I - \alpha\, hJ)^t\,, \qquad (9.6.7)$$

with the resultant MIRK formula only requiring t forward and backward substitutions. This approach requires $tm^2 + 1/3 m^3$ multiplications for a full matrix J in order to solve (9.6.4) and (9.6.7). For large IVPs (i.e., $m \gg t$), the computational effort of the MIRK is, thus, reduced to the same order as that of the BDFs.

Cash and Singhal (1982) derived the following L-stable methods of order three for the autonomous system $y' = f(y)$:

$$y_{n+1} = y_n + \frac{h}{6} \left[f(y_n) + 4f(\overline{y}_{n+\frac{1}{2}}) + f(y_{n+1}) \right]\,, \qquad (9.6.8a)$$

$$\overline{y}_{n+\frac{1}{2}} = \frac{1}{4}(3y_{n+1} + y_n) - \frac{h}{4} f(y_{n+1})\,, \qquad (9.6.8b)$$

whose iteration matrix

$$I - B = I - \frac{2}{3}hJ + \frac{1}{6}h^2J^2 = (I - \upsilon\,hJ)^2 \,, \tag{9.6.8c}$$

$$\upsilon = 0.385340199921340 \,.$$

The embedded L-stable MIRK scheme of order two is given as

$$y_{n+1} = y_n + h\left\{ \beta_0[f(y_n) + f(y_{n+1})] + \beta_1\,f(\bar{y}_{n+\frac{1}{2}}) \right\} \,, \tag{9.6.9}$$

with $\beta_0 = 0.1768828728131$, $\beta_1 = 0.646234254738$. The iteration matrix (9.6.8c) was also adopted.

10. SECOND ORDER DIFFERENTIAL EQUATIONS

10.1. Introduction

Gear (1967, 1971a,b) discussed the direct numerical solution of the m-th order differential system

$$y^{(m)} = f(x, y, y', ..., y^{(m-1)}), \tag{10.1.1}$$

with the tacit assumption that f is continuous in the independent variable x and satisfies a Lipschitz condition of order one with respect to y and each of $y^{(i)} = \dfrac{d^i y}{dx^i}, i = 1(1)m-1$.

Rutishauser (1960) examined the direct solution of (10.1.1) and its equivalent first order IVP (2.1.4) and concluded that the choice of approach depends on the particular problem at hand.

Henrici (1962) and Lambert (1973a) discussed the theory of the direct finite difference methods for the second order IVP

$$y'' = f(x, y, y'), y(a), y'(a) \text{ given} . \tag{10.1.2}$$

The application of the conventional numerical methods discussed in earlier chapters will demand for the reduction of (10.1.2) to a set of first order IVP

$$y' = z , y(a) = y_0 \tag{10.1.3}$$

$$z' = f(x, y, z) , z(a) = z_0 .$$

226

For the numerical solution of (10.1.2) by finite differences, we can either reduce (10.1.2) to a system of two first order equations and then apply standard methods for the first order systems, such as those treated in Chapters 4 and 5, or, alternatively, direct methods without reducing (10.1.2) to a system. Various particular step-by-step methods have been proposed in the literature, the most notable among these being the classical Runge-Kutta-Nystrom methods (Nystrom 1925)

$$y_{n+1} = y_n + h y'_n + h^2 \sum_{j=1}^{s} a_j Y_j ,$$

$$y'_{n+1} = y'_n + h \sum_{j=1}^{s} b_j Y_j ,$$

$$Y_i = f (x_n + \alpha_i h , y_n + \alpha_i h y'_n + h^2 \sum_{j=1}^{i-1} \beta_{ij} Y_j) . \qquad (10.1.4)$$

Hairer and Wanner (1976) have considered obtaining Nystrom type methods for the general second order differential equations (10.1.2), and they list order conditions for the determination of the parameters of the method. Complete families of Runge-Kutta-Nystrom methods of orders three and four can be found in Chawla and Sharma (1980, 1985).

Zurmuhl (1968), Henrici (1962), Gear (1971a), Chawla and Sharma (1980, 1981a,b, 1985), Cooper (1967), Hairer (1977), and Van der Houwen (1979) developed explicit and implicit Runge-Kutta-Nystrom methods for the numerical solution of (10.1.2).

Dormand and Prince (1987) proposed two classes, RICM4(3) and RICM6(4), of embedded Runge-Kutta-Nystrom formulas (10.1.5).

Mechanical systems without dissipation often lead to the IVP (10.1.2), with the first derivative y' not occurring explicitly on the r.h.s.; i.e.

$$y'' = f (x, y), y (a), y'(a) \text{ given} . \qquad (10.1.5)$$

Theoretical solutions of (10.1.5) are normally highly oscillatory. For the integration of special second order differential equation (10.1.5), Hairer (1977) obtained order conditions for Nystrom methods using certain tree-structures. Complete solutions to the order conditions and general families of Nystrom methods have been developed for methods of orders two to four in Chawla and Sharma (1981a) and fifth order methods in Chawla and Sharma (1981b).

Although the theory of numerical methods for (10.1.5) is closely related to that of (2.1.4), certain distinctions of stability and error estimator considerations still exist between the two categories of numerical methods.

Gautschi (1961), Stiefel and Betis (1969), Jain et al. (1979), and Sommeijer et al. (1986) proposed computational methods for Problem (10.1.5) whose solution period is known in advance. For the numerical solution of Problem (10.1.5) whose period is unknown, it is desirable that the numerical method be P-stable.

10.2. Linear Multistep Methods and the Concept of P-Stability

The general linear multistep method for the special second order initial value problem (10.1.5) is described by

$$\rho(E)y_n = h^2\sigma(E)f_n , \tag{10.2.1a}$$

where E is the shift operator such that

$$E^j y_n = y_{n+j} , \ j \text{ an integer} . \tag{10.2.1b}$$

ρ and σ are the first and second characteristic polynomials

$$\rho(E) = \sum_{j=0}^{k}\alpha_j E^j , \ \sigma(E) = \sum_{j=0}^{k}\beta_j E^j . \tag{10.2.1c}$$

Definition 10.1. A method is said to have error order p if

$$\rho(E)Z(x)-h^2 \ \sigma(E) \ Z''(x) = O(h^{p+2}) \tag{10.2.2}$$

for sufficiently smooth function $Z(x)$.

An important subclass of Method (10.2.1) is the Stormer-Cowell method, which is essentially (10.2.1) with

$$\rho(E) = E^{k-2}(E^2-2E+1) . \tag{10.2.3}$$

Stormer-Cowell formulas have identical properties as the Adams-Bashforth-Moulton formulas for first order IVPs (2.1.4). Stormer-Cowell formulas with stepnumbers greater than two are known to suffer from orbital instability (cf. Lambert and Watson 1976). For a test problem that describes uniform motion in a circular orbit, the numerical solutions generated by such methods spiral inwards for all values of steplength h.

Stieffel and Bettis (1969) suggested a modification of the Cowell methods for the integration of orbits. Their approach eliminated truncation errors and, thus, avoided the instability of the Cowell schemes.

Various other modifications of the Stormer-Cowell formulas have been proposed to eliminate this deficiency, but they all require an *a priori* knowledge of the frequency.

Lambert and Watson (1976) proposed an alternative strategy to eliminate the orbital instability that is inherent in the Stormer-Cowell methods. They suggested the adoption of the symmetric linear multistep methods, which are essentially (10.2.1) with the following constraints:

(a) α_j, β_j are real for $j = 0(1)k$, and

$$\alpha_k \neq 0, |\alpha_0| + |\beta_0| \neq 0 . \tag{10.2.4}$$

(b) ρ and σ are relatively prime; i.e.,

$$(\rho, \sigma) = 1 .$$ (10.2.5)

(c) Method (10.2.1) is consistent; i.e.,

$$\rho(1) = \rho'(1) = 0 , \quad \rho''(1) = 2\sigma(1) .$$ (10.2.6)

(d) Equation (10.2.1) is zero-stable; i.e., ρ has all its roots inside or on the unit circle, while those on the unit circle have multiplicity not greater than two.

(e) Equation (10.2.1) is symmetric; i.e.,

$$\alpha_j = \alpha_{k-j} , \; \beta_j = \beta_{k-j} , \; j = 0(1)\frac{k}{2} .$$ (10.2.7)

Conditions (10.2.5) and (10.2.7) imply that stepnumber k must be even and that order of method also must be even.

The stability/periodicity properties of the LMM (10.2.1) can be examined by its application to the scalar test problem

$$y'' + \lambda^2 y = 0 ,$$ (10.2.8)

which results in the finite difference equation

$$\sum_{r=0}^{k} Q_r(z^2) y_{n+r} = 0 , \; z = i\lambda h ,$$ (10.2.9)

where $Q_r(z^2)$ are polynomials in z^2.

The characteristic equation of the k-th order difference equation (10.2.9) is

$$\Pi(R, z) = \sum_{j=0}^{k} Q_j(z^2) R^j , \; k > 1 ,$$ (10.2.10)

which is expected to fulfill the following hypotheses (Hairer 1979):

$$Q_0(z^2) \neq 0 ,$$ (10.2.11)

$$\Pi(R, Z) \text{ is irreducible, and}$$ (10.2.12)

$$\Pi(1, 0) = \frac{\partial \Pi}{\partial R}(1, 0) = 0 , \text{ and,}$$ (10.2.13)

$$\frac{\partial^2 \Pi}{\partial R^2}(1, 0) \neq 0 .$$

Definition 10.2 (Lambert and Watson 1976). Method (10.2.1) is said to have an interval of periodicity $(0, H_0)$ if, for all $H = (\lambda h)^2$ in this interval, the roots R_i of the stability equation (10.2.9) satisfy $R_1 = e^{i\theta(H)} , R_2 = e^{-i\theta(H)} , |R_i| \leq 1 , i = 3(1)k$ for real $\theta(H)$.

Jeltsch (1978) gave an alternative definition.

Definition 10.3 (Lambert and Watson 1976). Method (10.2.1) is P-stable if its interval of periodicity is $(0, \infty)$.

Jeltsch (1978) has given a characterization for Linear Multistep methods possessing non-vanishing intervals of periodicity. Van der Houwen (1979) has analyzed a class of explicit Runge-Kutta-Nystrom methods of orders one and two for the integration of large systems of second order differential equations arising from the semi-discretization of certain classes of hyperbolic differential equations. A complete study of intervals of periodicity for classical Nystrom Methods can be found in Chawla and Sharma (1981a, b). In Chawla (1985b), it is shown that the stability properties of the Nystrom methods grow poor as their order increases, and increasing the number of stages in such methods do not really pay off by enhancing their stability properties. A new class of explicit substitution methods has been proposed in Chawla (1985b) that have the remarkable property that a method using $s+1$ stages has a scaled interval of periodicity (i.e., length of the interval of periodicity divided by $s+1$) that is very close to two. The cost of these methods is compensated by the relative increase in the length of the interval of periodicity.

Lambert and Watson (1976) proposed symmetric multistep methods of stepnumbers $k = 2, 4$, and 6, respectively, whose intervals of periodicity are non-vanishing. In fact, the two-step scheme is P-stable.

Definition 10.4 (Dahlquist 1978). Method (10.2.1) is said to be unconditionally stable, provided (10.2.10) satisfies the root condition.

Lambert and Watson (1976) and Dahlquist (1978) independently established the following barrier theorem as regards the attainable order of P-stable/unconditionally stable LMM (it must be mentioned that Lambert's complete proof was never published):

Theorem 10.1. *Method (10.2.1) is P-stable/unconditionally stable if*

(i)   it is implicit, i.e., $\beta_k \neq 0$, and

(ii) it is, at best, of order $p = 2$, and the most accurate P-stable/unconditionally stable LMM is

$$y_{n+2} - 2y_{n+1} + y_n = \frac{1}{4}h^2(f_{n+2}+2f_{n+1}+f_n) . \qquad (10.2.14)$$

Richtmyer and Morton (1967, p. 263) proposed (10.2.14) in connection with the solution of second order hyperbolic partial differential equations.

Equation (10.2.14) is equivalent to the implicit trapezoidal scheme (3.4.1) applied to the system $y\prime = z, z\prime = f(x, y)$. ρ, σ are perfect squares.

10.3. Derivation of P-Stable Formulas

The derivation of P-stable formulas is often based on the Padé approximation. Consider the polynomial $P_j(z)$ of degree j

$$P_j(z) = 1 + \frac{j}{2j}z + \frac{j(j-1)}{(2j)(2j-1)}\frac{z^2}{2!} + ... + \frac{(j!)(j!)}{2j!}\frac{z^j}{j!} . \tag{10.3.1}$$

Then $P_s(z)/P_s(-z)$ is the (s, s) Padé approximation to $\exp(z)$ and

$$Q(s, R, z) = (P_s(-z)R - P_s(z))(P_s(z)R - P_s(-z)) \tag{10.3.2}$$

is of the form (10.2.9).

Butcher (1965), Gear (1965), and Gragg and Stetter (1964) suggested the use of hybrid methods in order to avoid the stability and convergence constraints imposed by the Dahlquist theorems (1956, 1963) on the linear multistep formulas (5.1.1) for first order IVPs (2.1.4). Gupta (1979) discussed the polynomial representation of such methods, as well as their implementation, using the Nordsieck vector representation.

Chawla (1981) and Cash (1981) independently showed that the "order barrier" imposed by Theorem 10.1 on the attainable order of P-stable methods could be crossed-over by considering certain hybrid two-step methods. They also showed the existence of fourth and sixth order P-stable methods. General families of two-step fourth order P-stable methods are given in Chawla and Neta (1986), where their implementation for nonlinear problems and their efficiency are examined. Thomas (1984), Chawla and Rao (1985, 1986), Van der Houwen (1979), and Sommeijer et al. (1986) developed methods along this line.

Hairer (1979) used the concept of Padé approximation to obtain the following P-stable hybrid scheme of order $p = 4$, using one offpoint:

$$y_{n+2} - 2y_{n+1} + y_n = \frac{1}{12}h^2\{f(x_{n+2}, y_{n+2})$$

$$+ (10 - \gamma)f(x_{n+1}, y_{n+1}) \tag{10.3.3a}$$

$$+ f(x_n, y_n) + \gamma f(x_{n+1}, \bar{y}_n)\} ,$$

$$\bar{y}_n = y_{n+1} - \frac{1}{12\gamma}h^2\{f(x_{n+2}, y_{n+2})$$

$$\tag{10.3.3b}$$

$$- 2f(x_{n+1}, y_{n+1}) + f(x_n, y_n)\} ,$$

provided

$$\gamma \neq 0 , (\lambda h)^2 - 12 \neq 0 . \tag{10.3.3c}$$

Hairer's formula (10.3.3) is stabilized, except at $\lambda h = \sqrt{12}$. A P-stabilized (unconditionally stabilized) method is given by Chawla (1983).

10.4. One-Leg Multistep and One-Leg Hybrid Methods for Second Order IVPs

In view of the significant advantages enjoyed by the one-leg multistep methods with the first order IVPs (10.1.3), as discussed in Section 8.7, Fatunla (1984, 1985a,b,c) considered the extension of this concept to develop P-stable one-leg multistep and one-leg hybrid formulas for the numerical solution of second order IVPs (10.1.5).

Fatunla (1984) used the idea of (1, 1) Padé approximation to realize the one-leg twin of (10.2.14); i.e.,

$$y_{n+2} - 2y_{n+1} + y_n = h^2 f\left(x_{n+1}, \frac{1}{4}(y_{n+2} + 2y_{n+1} + y_n)\right) . \qquad (10.4.1)$$

In order to avoid the Dahlquist/Lambert Barrier Theorem (Theorem 10.1), Fatunla (1984b) considered the following one-leg hybrid method:

$$\alpha_2 y_{n+2} + \alpha_1 y_{n+1} + \alpha_0 y_n = h^2 f\left(\theta_n, \gamma y_{n+v} + \sum_{j=0}^{2} \beta_j y_{n+j}\right), \qquad (10.4.2a)$$

$$y_{n+v} + \alpha'_1 y_{n+1} + \alpha'_2 y_{n+2} = h^2 f\left(\theta'_n, \sum_{j=0}^{2} \beta'_j y_{n+j}\right), \qquad (10.4.2b)$$

$$\theta_n = \gamma x_{n+v} + \sum_{j=0}^{2} \beta_j x_{n+j}, \qquad (10.4.2c)$$

$$\theta'_n = \sum_{j=0}^{2} \beta'_j x_{n+j}, \qquad (10.4.2d)$$

$$\sum_{i=0}^{2} \beta'_j = 1, \ \sum_{j=0}^{2} \beta_j + \gamma = 1. \qquad (10.4.2e)$$

On applying (10.4.2) to the scalar test problem (10.2.8), we obtain a second order difference equation (10.2.9) with the following characteristic polynomial:

$$\Pi(R, z) = |1-(\beta_2-\gamma\alpha_2)z^2 + \gamma\beta'_2 z^4| R^2$$

$$- |2+(\beta_1-\gamma\alpha_1)z^2 + \gamma\beta'_1 z^4| R \qquad (10.4.3)$$

$$+ |1+\beta_0 z^2 + \gamma\beta'_0 z^4| .$$

By expecting (10.4.3) to coincide with $Q(2, R, z)$, the coefficients of the one-leg hybrid multistep scheme (10.4.2) are

$$\alpha_2 = \alpha_0 = 1 , \ \alpha_1 = -2 ,$$

$$\beta_0 = \frac{1}{12} \ , \ \beta_1 = \frac{5}{6} + \frac{1}{36}(r-2) \ ,$$

$$\beta_2 = \frac{1}{12} + \frac{1}{36}(1-r) \ , \ \gamma = \frac{1}{36} \ ,$$

$$\alpha_1' = r-2 \ , \ \alpha_2' = 1-r \ , \text{ and}$$

$$\beta_0' = \frac{1}{4} = \beta_2' \ , \ \beta_1' = \frac{1}{2} \ . \tag{10.4.4}$$

Some striking examples of the one-leg multistep scheme (10.4.2) include

$$y_{n+2} - 2y_{n+1} + y_n = h^2 f \left(\theta_n, \frac{1}{9}y_{n+2} + \frac{14}{18}y_{n+1} + \frac{1}{5}y_n\right) \tag{10.4.5}$$

and

$$y_{n+2} - 2y_{n+1} + y_n = h^2 f \left(\theta_n, \frac{1}{12}y_{n+2} + \frac{5}{6}y_{n+1} + \frac{1}{12}y_n\right) \ . \tag{10.4.6}$$

The order conditions for the one-leg multistep method

$$\rho(E)y_n = h^2 f \left(\sigma(E)y_n\right) \ , \ \sigma(1) = 1 \tag{10.4.7}$$

of order $p = \min\{q, s\}$, in case of the homogeneous second order differential system

$$y'' = f(y) \ , \tag{10.4.8}$$

are specified as follows:

$$\sum_{j=0}^{k} \alpha_j = 0 \ , \ \sum_{j=0}^{k} j\alpha_j = 0 \ , \tag{10.4.9a}$$

$$\sum_{j=0}^{k} \alpha_j \gamma_j^r = r(r-1)\bar{h}^{-2} \sum_{j=0}^{k} \beta_j \gamma_j^{r-2} \ , \ r = 2(1)q+1 \ , \tag{10.4.9b}$$

$$\sum_{j=0}^{k} \beta_j \gamma_j^{r-2} = \left[\sum_{j=0}^{k} \beta_j \gamma_j\right]^{r-2} \ , \ r = 2(1)s+1 \ . \tag{10.4.9c}$$

Conditions (10.4.9a-c) are not required by the conventional LMM (10.2.1). P-stability, like A-stability, imposes implicitness on such methods. However, in view of the very large systems that one encounters when, for example, hyperbolic equations are semi-discretized, it may be advantageous to use explicit methods in order to reduce storage requirements. The disadvantages of a restricted maximal integration step is partly compensated by the arrival of vector computers, on which explicit methods can be implemented with great efficiency. Particularly useful in this direction will be the new class of substitution methods proposed by Chawla (1985a,b), which are explicit, and a

Table 10.1. Coefficients, Error constants, and Interval of Periodicity of the Multiderivative Methods (10.5.5) (Twizell and Khaliq 1984)

r	s	α_1	α_2	α_3	α_4	β_1	β_2	β_3	β_4	c_{r+s+v}	Interval of Periodicity
0	1	0	0	0	0	0	0	0	0	1	*
1	1	$\frac{-1}{4}$	0	0	0	$\frac{1}{2}$	0	0	0	$\frac{-1}{6}$	(θ, ∞)
1	0	-1	0	0	0	0	0	0	0	-1	*
0	2	0	0	0	0	1	0	0	0	$\frac{1}{12}$	$(0, 4)$
1	2	$\frac{-1}{9}$	0	0	0	$\frac{7}{9}$	0	0	0	$\frac{-1}{36}$	$(0, \frac{36}{5})$
2	2	$\frac{-1}{12}$	$\frac{1}{144}$	0	0	$\frac{5}{6}$	$\frac{1}{72}$	0	0	$\frac{1}{360}$	$(0, \infty)$
2	1	$\frac{-1}{9}$	$\frac{1}{36}$	0	0	$\frac{7}{9}$	0	0	0	$\frac{1}{36}$	$(0, \infty)$
2	0	0	$\frac{1}{4}$	0	0	1	0	0	0	$\frac{7}{12}$	$(0, \infty)$
0	3	0	0	0	0	1	0	0	0	$\frac{1}{12}$	$(0, 4)$

Table 10.1 continued

										Interval
1	3	$-\frac{1}{16}$	0	0	0	$\frac{7}{8}$	$\frac{1}{48}$	0	0	$\frac{-7}{2880}$	$(0, 6.5)\alpha$ $(29.5, 48)$
2	3	$-\frac{3}{50}$	$\frac{1}{400}$	0	0	$\frac{22}{25}$	$\frac{17}{600}$	0	0	$\frac{1}{3600}$	$(0, 8.2)\alpha$ $(14.6, \frac{300}{7})$
3	3	$-\frac{1}{120}$	$\frac{1}{600}$	$\frac{-1}{14400}$	0	$\frac{9}{20}$	$\frac{11}{300}$	$\frac{1}{7200}$	0	$\frac{-1}{50400}$	$(0, \infty)$
3	2	$-\frac{3}{50}$	$\frac{1}{400}$	$\frac{-1}{3600}$	0	$\frac{22}{25}$	$\frac{17}{600}$	0	0	$\frac{-1}{3600}$	$(0, \infty)$
3	1	$-\frac{1}{16}$	0	$\frac{-1}{576}$	0	$\frac{7}{8}$	$\frac{1}{48}$	0	0	$\frac{-17}{2880}$	$(0, \infty)$
3	0	$-\frac{1}{12}$	0	$\frac{-1}{36}$	0	1	0	0	0	$\frac{-1}{12}$	$(0, \infty)$
0	4	0	0	0	0	1	$\frac{1}{12}$	0	0	$\frac{1}{360}$	$(0, 12)$

*Inconsistent Scheme.

$$v = \begin{cases} 1 & \text{if } r + s \text{ is odd,} \\ 2 & \text{if } r + s \text{ is even.} \end{cases}$$

method using $(s+1)$-stages that possesses an interval of periodicity of size very nearly $2(s+1)$.

The one-stage second order Runge-Kutta Formula is unconditionally stable in the sense of Dahlquist, and Hammer and Hollingsworth; i.e., Equation (4.6.8) is stable except for the single point $H = \sqrt{12}$.

It is to be hoped that type-insensitive P-stable codes will soon be available for the second order IVP (10.1.5).

10.5. Multiderivative Methods for Second Order IVPs

Brown (1974), Jeltsch (1976), Lambert (1973a), Enright (1974a), and Twizell and Khaliq (1981, 1984) established that multidervative methods give high accuracy and possess good stability properties when used to solve first order IVPs (2.1.4).

The good accuracy/stability properties of multiderivative methods, of course, follow from the order star theory in the proof of the Daniel-Moore conjecture by Wanner et al. (1978).

Twizell and Khaliq (1984) proposed a class of P-stable two-step higher derivative formulas for the special second order IVPs (10.1.5).

Their method was based on the fact that the theoretical solution

$$y(x) = Ae^{i\lambda h} + Be^{-i\lambda h}, \ i^2 = -1, \ A^2 + B^2 \neq 0 \tag{10.5.1}$$

to the scalar test problem (10.2.8) satisfies the following recurrence relation:

$$y(x+h) - \left[e^{i\lambda h} + e^{-i\lambda h} \right] y(x) + y(x-h) = 0. \tag{10.5.2}$$

They adopted the (r,s) Padé approximant

$$e^{\alpha h} = \frac{P_r(\alpha h)}{Q_s(\alpha h)} + O(h^{r+s+1}), \tag{10.5.3a}$$

where

$$P_r = \sum_{j=0}^{r} a_j x^j, \ Q_s = \sum_{j=0}^{s} b_j x^j. \tag{10.5.3b}$$

The insertion of successive derivatives of (10.2.7) in (10.5.2) yields the multiderivative integration formula

$$y_{n+1} + \sum_{j=1}^{r} \alpha_j h^{2j} y_{n+1}^{(2j)} = 2 y_n + \sum_{\mu=1}^{k} \beta_\mu h^{2\mu} y_n^{(2\mu)}$$

$$\tag{10.5.4a}$$

$$- y_{n-1} - \sum_{j=1}^{r} \alpha_j h^{2j} y_{n-1}^{(2j)} ,$$

$$k = [½ (r+s)]. \tag{10.5.4b}$$

The perodicity equation $\Pi(R, H^2)$ for the multiderivative formula (10.5.4) can be expressed in terms of the Padé approximant as

$$\Pi(R, H^2) = Q_s(-iH) Q_s(iH) R^2 - (Q_s(-iH)P_r(iH)$$

$$+ Q_s(iH) P_r(-iH))R + Q_s(-iH)Q_s(iH) = 0 . \tag{10.5.5}$$

The interval of periodicity of (10.5.4) is obtained by computing the values of H^2 for which R_1, R_2 satisfy

$$R_1 = e^{i\,\theta(H)}, R_2 = e^{-i\,\theta(H)} , \tag{10.5.6}$$

where $\theta(H)$ 1s a real number.

Twizell and Khaliq (1984) reproduced the non-zero coefficients a_j, b_j for the first sixteen entries—the error constants as well as the intervals of periodicity—of the Padé Table for the exponential function. These are given in Table 10.1.

They also implemented the multiderivative method (10.5.4) in a PECE mode, using a general (r,s) Padé approximant as predictor and a general (r,s) Padé approximant as corrector, with possible adoption of Milne's error estimate if $s^* = r+s$.

Goyal and Serbin (1987) employed the theory of Butcher series (1964, 1972) to develop a class of generalized Rosenbrock-type methods for (10.1.5).

It is hoped that type-insensitive P-stable codes will soon be developed for the class of second order IVP (10.1.5).

11. RECENT DEVELOPMENTS IN ODE SOLVERS

Gear (1969, 1971a,b) produced the code DIFSUB, which combines the Adams-Moulton method for a nonstiff step and the backward differentiation scheme for a stiff step. This code has been subjected to various refinements and modifications, culminating in the most robust code LSODE and its variants (Hindmarsh 1980). The chart of development from DIFSUB to LSODE is given in Table 9.4.

Enright et al. (1975) observed that DIFSUB is unrivalled by codes based on implicit Runge-Kutta processes, second derivative methods (Enright 1974a,b), and extrapolation processes (Lindberg 1973) if the eigenvalues of the Jacobian are not close to the imaginary axes (i.e., stiff oscillatory problems). Codes based on these other methods have since been considerably improved upon, as their efficiency is now close to that of LSODE.

This chapter examines some of the innovations which have been built into various ODE solvers to improve their efficiency, reliability, and robustness. Considerable improvements have been attained in error estimation, and the expertise of ODE solvers is being extended to differential/algebraic equations, boundary value problems, and partial differential equations.

With the Dahlquist Barrier Theorem (Theorem 3.3) (1963), all conventional ODE solvers for stiff IVPs are implicit, with the result that a good fraction of computational effort is expended on solving linear algebra involved in the solution of the nonlinear systems arising from the implicitness of presently-available stiff algorithms.

The matrix algebra involved in the adoption of implicit stiff algorithms includes

(i)   the evaluation of Jacobian,

(ii) LU decomposition of the iteration matrix, and

(iii) backward substitution.

Efforts aimed at a drastic reduction of the linear algebra in stiff algorithms are

(a) Infrequent evaluation of the Jacobian (cf. Equations (8.3.11 and (8.3.12)) by adopting the modified Newton iteration.

(b) Transformation of the iteration matrix into a more efficient form; e.g., Enright (1974a,b) introduced the Constraint (9.4.5) so as to be able to factorize the iteration matrix (9.4.4) of the second derivative formula (9.4.3) into a perfect square, as given by (9.4.7). This procedure (i) reduces the computational efforts by $O(m^3)$ since there is no need to compute J^2 again, and (ii) it permits the exploitation of the sparsity of the Jacobian J. Enright and Kamel (1980) obtained similar results for the blended DIFSUB (cf. Skeel and Kong 1977).

(c) The introduction of (i) the semi-implicit Runge-Kutta scheme (Norsett 1974) (uncoupled nonlinear systems), (ii) the diagonally implicit Runge-Kutta schemes (Alexander 1977), (iii) the Rosenbrock methods (Rosenbrock 1963, Kaps and Wanner 1981, Shampine 1982), and (iv) the singly implicit Runge-Kutta methods (Burrage 1978) has considerably reduced the amount of linear algebra associated with the implicit Runge-Kutta methods (uncoupled nonlinear systems). The resultant codes— SIRK, DIRK, STRIDE, SIMPLE—are almost as efficient as those based on BDF, and they enjoy better stability characteristics. Dormand and Prince (1980) and Prince and Dormand (1981) proposed efficient DIRK methods whose accommodation of interpolation (cf. Horn 1981, Shampine 1982, 1985c) renders it particularly useful as an automatic code, and easily adaptable to handle discontinuities (cf. Enright et al. 1986). The mono-implicit R-K methods (Cash 1975, 1977a,b; Bokhoven 1980; Cash and Singhal 1982) require an amount of $O(m^3)$ operations equivalent to the BDF.

(d) The efficiency of the various type of Runge-Kutta formulas has greatly improved since the idea of embedding was introduced to obtain local truncation errors (Fehlberg 1969, Burrage 1978).

(e) Bader and Deuflhard (1983) developed the code METAN1 based on the semi-implicit midpoint rule combined with polynomial extrapolation. Apart from established superior stability properties over the BDF, it compares favorably in terms of efficiency with those codes based on BDF. Shampine (1983c) updated this into a type-insensitive code METAN2. The error estimation strategy suggested by Deuflhard (1983, 1985) has significantly increased the efficiency of the extrapolation methods.

(f) Another bold attempt at reducing the matrix linear algebra of stiff ODE solvers is automatically partitioning the stiff differential system and exploiting the resulting structure. There are two basic approaches:

(i)   As early as 1952, Dahlquist suggested that the differential system (11.1.4) be partitioned into stiff and nonstiff (transient and smooth) components, and that one then use some efficient technique to reduce the cost of iteration. This concept is similar to that of singular perturbation whereby the IVP

$$y' = f(y(x)), \ y(0) = y_0 \tag{11.1.1}$$

is transformed into

$$\begin{bmatrix} Du' \\ v' \end{bmatrix} = \begin{bmatrix} f_1(u, v) \\ f_2(u, v) \end{bmatrix}, \tag{11.1.2}$$

where

$$y = \begin{bmatrix} u \\ v \end{bmatrix}, \ f(y) = \begin{bmatrix} f_1(u, v) \\ f_2(u, v) \end{bmatrix}, \tag{11.1.3}$$

and where D is a matrix which is possibly singular, u corresponds to the transient components, and v corresponds to the smooth components.

Dahlquist (1952) proposed the idea of *pseudostationary* approximation (PSA) by setting $D = 0$, thus reducing (11.1.2) to a differential algebraic equation (DAE)

$$F(x, y, y') = 0. \tag{11.1.4}$$

Petzold (1982a,b; 1984) and Gear and Petzold (1984) suggested the numerical solution of the DAE with stiff codes whose error estimators and other strategies have to be extensively modified, with the added risk that stepsize adjustments may cause large errors.

The first order BDF is the simplest method to solve the DAE (11.1.4) leading to the nonlinear equation

$$F(x_{n+1}, y_{n+1}, \frac{(y_{n+1} - y_n)}{(x_{n+1} - x_n)}) = 0. \tag{11.1.5}$$

The BDF, RKF, and Extrapolation methods are considered generalizations of this idea. For instance, the k-step BDF expressed as

$$y_{n+1}' = \frac{1}{h} \sum_{j=0}^{k} \alpha_j \, y_{n+1-j}$$

can be applied directly to Problem (11.1.4).

In the event that $\partial F / \partial y'$ is nonsingular, (11.1.4) can be transformed to the IVP (2.1.4) at a cost of forfeiting the sparsity of the original system and the requirement for more core storage.

Gupta et al. (1985) discussed the organization of an automatic code for DAE and proposed a code which can handle a reasonable class of problems with index ≤ 2.

Existing algorithms to cope with DAE (11.1.4) include those of Gear (1971c), Sincovec et al. (1981), Starner (1976), Petzold (1981, 1982a,b, 1984, 1986), Petzold and Gear (1982), Gear and Petzold (1984), Berzins and Dew (1986), Brenan (1986), and Leimkuhler (1986).

In the last few years, the efficiency of existing ODE codes has been largely increased with the introduction of the concept of partitioning.

Some authors suggest the use of an analytic expansion method or a stiff algorithm for the stiff sub-system in (11.1.2) while the nonstiff sub-system is solved with a nonstiff method (MacMillan 1968, Finden 1975, Lee 1967, Hoffer 1976, Alfeld and Lambert 1977).

However, the accuracy of this approach is very sensitive to correct partitioning and the amount of coupling existing between the transient and smooth subsystems. Apart from this, the partitioning (11.1.2) is time-dependent.

(ii) Robertson (1976), Carver and MacEwen (1981), Enright and Kamel (1981), and Krogh (1982) applied implicit integration formulas to all the components of the differential system (11.1.1) and exploited partitioning only in the iterative solution of the resulting linear systems of equations. This is equivalent to adopting a modified Newton for the transient component and a functional iteration for the smooth component.

The advantage of this approach is that only the cost of the iteration will be sensitive to the correct identification of the transient and smooth subsystems, while the accuracy of the numerical results will be insensitive to the partitioning. Kamel (1978) and Enright and Kamel (1980) have successfully applied the automatic partitioning to increase the efficiency of

(i)   the linear multistep formulas,

(ii) the second derivative formulas,

(iii) the blended DIFSUB, and

(iv) Runge-Kutta methods.

Gear (1980a,d), Watts (1983), Gladwell et al. (1987), and Shampine (1987) revisited the issue of automatic selection of initial stepsize h_0 for ODE codes.

The approach suggested by Shampine (1987) is practical, cheap, efficient, reliable, and applicable to both the Adams and Backward Differentiation formulas. It also eliminates the mesh distortion problem associated with Sedgwick's (1973) method of taking the smallest permissible machine unit as the initial stepsize and then building up both the order and stepsize very rapidly for the Adams method.

The original predictor-corrector implementation of the Adams method in Shampine and Gordon (1975) is to generate two numerical solutions y_1 and y_1^*

at $x_1 = x_0 + h$, adopting the explicit Euler and the trapezoidal scheme, respectively, with h just small enough to satisfy the condition

$$y_1^* - y_1 = \frac{1}{2} hf_1 - f_0 \leq TOL ,\qquad (11.1.6)$$

where TOL is the error tolerance.

Gear (1971a,b) in the code DIFSUB, and Hindmarsh (1980) in the codes LSODE and LSODI, adopted the same approach but with y_1 generated using the implicit Euler to start the Backward Differentiation formula. Shampine (1987) felt that there is no need to use implicit Euler since it is more expensive to implement and it has asymptotically the same local truncation error as the explicit Euler.

In order to guarantee the credibility of the initial meshsize, Shampine (1987) advocated that the trial initial stepsize, in addition to satisfying (11.1.6), should also satisfy the more stringent constraint

$$\frac{1}{2} h[f_1 + f_0] \leq TOL ,\qquad (11.1.7)$$

and that a trial stepsize

$$h = \frac{1}{5} TOL * f_0^{-1}\qquad (11.1.8)$$

be used to generate y_1 and y_1^* provided that $f_0 \neq 0$.

With an initial stepsize which satisfies Equations (11.1.6) and (11.1.7), an optimal stepsize

$$h_{opt} = (TOL * y_1^* - y_1^{-1}) * h\qquad (11.1.9)$$

is the largest stepsize for which all $h \leq h_{opt}$ satisfy (11.1.6) provided the asymptotic behavior of the formula is evident.

Another development in ODE solvers is in the area of improving their efficiency in the neighborhood of discontinuities that can be located if switching functions were provided. Gear and Østerby (1984) used the local error estimates to locate and detect discontinuities and incorporated order-finding techniques as well as efficient methods to restart beyond the point of discontinuity. Carver (1977, 1978), in the code FORSIM, incorporated the discontinuity function into the differential system, while Hindmarsh (1979, 1983a,b), in the code LSODAR, built in a root-finder to locate the point of discontinuity. Shampine and Gordon (1975), in the code STEP, made provision to deal with discontinuities. Enright et al. (1986) modified the R-K methods of Dormand and Prince (1980) to handle discontinuities.

Petzold and Gear (1977) and Petzold (1978, 1981) used the idea of a quasi-envelope to transform a differential equation that involves the integral of the original, highly oscillatory problem into a smooth function that can be integrated using a large (possibly larger than the period of oscillation) meshsize. This approach is an extension of the multi-revolution methods used by astronomers in the calculation of the orbits of artificial satellites (cf. Mace and Thomas 1960, Graff 1973, Graff and Bettis 1975).

Also, serious attention has been given to cyclic methods, which, because of their superior stability, perform better on stiff oscillatory problems than codes based on the BDF (Bickart et al. 1971, 1973; Albrecht 1985, 1987; Iserles 1984; Tischer and Sacks-Davis 1983; Tendler et al. 1978; Tischer and Gupta 1983).

Albrecht (1987) proposed a new theoretical approach to R-K methods in the spirit of A-methods (cf. Albrecht 1978, 1985, 1986, 1987). This approach

(i)   eliminates the traditional difference in the treatment of R-K methods and Linear Multistep methods, and also

(ii)   eliminates the need to consider elementary differentials.

Gear (1987) examined several potential areas of application of parallelism across the ODE systems and across ODE methods. Major impediments to parallelism in ODEs include

(i)   the narrowness of the computational lattice in most ODE algorithms, and

(ii)   the problem of communicating information between or through processes.

Unfortunately, stiff ODE codes are normally implicit, require the solution of nonlinear systems of equations, and, thus, require a complete intercommunication between processes, although the iteration matrix W is reducible.

Duff et al. (1985) established the following theorem to serve as a guide:

Theorem 11.1. *W irreducible implies W^{-1} is structurally full.*

In addition, the most interesting problems in ODEs are not reducible because of the coupling between components.

Deuflhard (1985) argued that extrapolation methods can be executed in parallel by using separate processors to compute each unextrapolated approximation.

Gear (1987) suggested that each internal approximation of an s–stage IRK method be solved on different processor, using the functional iteration

$$z_i^{(r)} = y_n + \sum_{j=1}^{s} a_{ij} Y_j^{(r)} , \qquad (11.1.10a)$$

$$Y_i^{(r+1)} = hf\left(z_i^{(r)}\right) . \qquad (11.1.10b)$$

Gear (1987) offered some solutions to these impediments and raised some open questions. Chu and Hamilton (1986, 1987), Norsett (1986), Gear and Wang (1987), also examined possible approaches to the parallel processing of numerical ODE methods. The application of parallelism to multi-block and R-K methods are discussed. Franklin (1978) gave a review of parallel solution of

ODEs while Ortega and Voigt (1985) gave a comprehensive survey on the application of parallelism to PDEs.

In the last few years, increasing attention has been given to the estimation of global errors in ODE codes.

Zadunaisky (1976) proposed an ingenious method to generate global errors by formulating a Neighboring Problem (NP)

$$z\prime = f(x, z) + d(x), \; z(a) = y_0 \qquad (11.1.11a)$$

with

$$d(x) = P'_k(x) - f(x, P_k(x)) \qquad (11.1.11b)$$

for the Original Problem (OP) (2.1.4).

(11.1.11) is an artificial problem in that its exact solution $z(x) = P_k(x)$ is a piecewise polynomial of degree $\leq k$ interpolating the numerical solution y_n to the OP (2.1.4).

The corresponding numerical solution $\{z_n\}$ to (11.1.11) can be generated by the same numerical method used to generate $\{y_n\}$.

The global error for the solution to OP (2.1.4)

$$E_n = y_n - y(x_n) \qquad (11.1.12)$$

is estimated by the global error for the NP (11.1.11)

$$W_n = z_n - P_k(x_n) . \qquad (11.1.13)$$

Worth noting are the excellent survey papers of Stetter (1978), Prothero (1980), Bohmer and Stetter (1984), and Skeel (1986a,c).

Skeel (1986c) classified global error estimation techniques into five distinct categories:

(1)   Defect Correction: Difference Correction, Zadunaisky's Neighboring Problem Idea, Solving for the Correction, Newton's Method.

(2)   Integrating the Principal Error Equation.

(3)   Richardson Extrapolation.

(4)   Using Two Different Tolerances.

(5)   Using Two Different Methods.

These techniques all require at least two parallel integrations, and of particular interest is the Defect Correction approach—a powerful tool normally adopted to increase the accuracy of solution.

The three basic steps in the defect correction approach for any given differential system and a particular numerical integrator are:

(a)   Obtain the defect of the numerical solution as a measure of how well the given problem has been solved,

(b) Use this defect in a simplified version of the problem to obtain the appropriate correction quantity. This step normally involves the computation of an approximate Jacobian either for the continuous error equation

$$e\prime(x) = f_y(x, \, y(x)) \, e(x) + d(x) \,, \tag{11.1.14}$$

with the global error given as

$$E(x_n) = h^p e(x_n) \,, \tag{11.1.15}$$

or in the discrete error equation for the Linear Multistep Method

$$E_n = \sum_{j=1}^{k-1} \alpha_j \, E_{n-j}$$

$$\tag{11.1.16}$$

$$+ h_n \sum_{j=0}^{k} \beta_j \, f_y(x_{n-j}, \, y(x_{n-j})) E_{n-j} + d_n$$

(c) Apply the correction to the approximate solution to give improved values.

In addition, vigorous effort to exploit the current expertise in ODE solvers of the numerical solution of two point Boundary Value Problems (BVPs), whose current state can be described as primitives is desirable. Enright (1985) gave an overview of numerical methods for two-point BVPs and identified strategies and approaches that could significantly improve the existing methods based on some useful codes:

(i) Multiple Shooting Methods (BVPSOL, Deuflhard and Bader 1982),

(ii) Collocation Methods (PASVA3, Lentini and Pereyra 1979),

(iii) Finite Difference Methods (COLSYS. Asher, Christiansen and Russell 1981).

At an abstract level, the three categories of methods for BVPs share a common structure—using a modified Newton iteration to obtain numerical approximation, which is determined by a discrete solution Y satisfying the nonlinear equations $g(Y) = 0$. Within the next decade, robust codes for two-point BVPs should be within the reach of both the casual and sophisticated users.

It is hoped that the following years will witness intense research activities to improve the efficiency of ODE codes on large stiff and stiff oscillatory initial value problems, as well as the development of efficient low-accuracy methods for large stiff initial value problems derived from the application of method of lines to parabolic, partial differential equations. Hindmarsh (1985) discussed existing methods for large stiff ODE systems. Iserles (1981) proposed a family of two-step A-stable methods for parabolic PDEs. Salane (1986) developed a package DEBDFs to handle large sparse systems of stiff ODEs.

REFERENCES

Addison, C. A. (1979), "Implementing a Stiff Method Based Upon the Second Derivative Formula," *Technical Report No.* **130**, Department of Computer Science, University of Toronto, Canada.

Addison, C. A. (1980), "Results of the 1979 ODE Olympics," Unpublished Manuscript.

Addison, C. A. (1984), "A Comparison of Several Formulas on Lightly Damped, Oscillatory Problems," *SIAM Journal on Scientific and Statistical Computing* **5**, 920-936.

Addison, C. A., and Gladwell, I. (1984), "Second Derivative Methods Applied to Implicit First- and Second-Order Systems," *International Journal for Numerical Methods in Engineering* **20**, 1211-1231.

Aitken, A. C. (1932), "On Interpolation by Iteration of Proportional Parts," *Proceedings Edinburgh Mathematical Society* **2**, 56-76.

Aiken, R.C. (1985), "Stiff Computation," Proceedings of the International Conference on Stiff Computation (R. Aiken, ed.), (April 1982; Park City, UT), New York: Oxford Press.

Albrecht, P. (1978a), "Explicit, Optimal Stability Functionals and Their Applications to Cyclic Discretization Methods," *Computing* **19**, 233-249.

Albrecht, P. (1978b), "On the Order of Composite Multistep Methods for Ordinary Differential Equations," *Numerische Mathematik* **29**, 381-396.

Albrecht, P. (1985), "Numerical Treatment of ODEs: The Theory of *A*-Methods," *Numerische Mathematik* **47**, 59-87.

Albrecht, P. (1986), "Numerical Treatment of Ordinary Differential Equations: The Theory of A-Methods," ODE Conference held at Sandia National Lab., Albuquerque, New Mexico, USA, July 1986.

Albrecht, P. (1987), "A New Theoretical Approach to Runge-Kutta Methods," *SIAM Journal on Numerical Analysis* **24**, 391-406.

Alexander, R. (1977), "Diagonally Implicit R-K Methods for Stiff Ordinary Differential Equations," *SIAM Journal on Numerical Analysis* **14**, 1006-1021.

Alexander, R. (1986), "A Type-insensitive ODE Code Based on DIRK Formulas," ODE Conference held at Sandia National Lab., Albuquerque, New Mexico, USA, July 1986.

Alfeld, P. (1986), "Quasi-Newton Methods for Stiff Ordinary IVPs," Technical Report, Department of Mathematics, University of Utah, Salt Lake City, Utah, USA.

Alfeld, P. S., and Lambert, J. D. (1977), "Correction in the Dominant Spaces: A Numerical Technique for a Certain Class of Stiff IVPs," *Mathematics of Computation*, **31**, 922-938.

Allen, R. H., and Prottle, C. (1966), "Stable Integration Methods for Electronic Circuit Analysis with Widely Separated Time Constants," *Proceedings 6th Annual Allerton Conference on Circuit and System Theory*, (T. Trick and R.T. Chien, eds.), University of Illinois at Urbana-Champaign, 311-320.

Amdursky, V., and Ziv, A. (1974), "On the Numerical Treatment of Stiff, Highly Oscillatory Systems," *IBM Technical Report* **015**, IBM Israel Scientific Center

Amdursky, V., and Ziv, A. (1975), "On the Numerical Solution of Stiff Linear Systems of the Oscillatory Type," *IBM Technical Report* **032**, IBM Israel Scientific Center.

Amdursky, V., and Ziv, A. (1976), "The Numerical Treatment of Linear Highly Oscillatory ODE Systems by Reduction to Non Oscillatory Types," *IBM Technical Report* **039**, IBM Israel Scientific Center.

Apostol, T.M. (1957), *Mathematical Analysis, A Modern Approach to Advanced Calculus*, Massachusetts: Addison-Wesley.

Ascher, U., Christiansen, J., and Russel, R. D. (1981), "Collocation Software for Boundary Value ODEs," *ACM Transactions on Mathematical Software* **7**, 209-229.

Atkinson, K. E. (1978), *An Introduction to Numerical Analysis*, New York: John Wiley & Sons.

Axelsson, O. (1969), "A Class of A-stable Methods," *BIT* **9**, 185-199.

Axelsson, A.O.H., and Verwer, J. G. (1985), "Boundary Value Techniques for IVPs in ODEs," *Mathematics of Computation* **45**, 153-172.

Bader, G., and Deuflhard, P. (1983), "A Semi-Implicit Midpoint Rule for Stiff Systems of ODEs," *Numerische Mathematik* **41**, 373-398.

Barnett, S. (1975), *Introduction to Mathematical Control Theory*, Oxford: Clarendon Press.

Bartle, R. G. (1975), The Elements of Real Analysis, New York: John Wiley & Sons.

Barton, D. (1980), "On Taylor Series and Stiff Equations," *ACM Transactions on Mathematical Software* **6**, 280-294.

Barton, D., Willers, I. M., and Zahar, R.V.M. (1971a) "The Automatic Solution of Ordinary Differential Equations by the Method of Taylor," *Computer Journal* 14, 243-348.

Barton, D., Willers, I. M., and Zahar, R.V.M. (1971b), "Taylor Series Methods for Ordinary Differential Equations—An Evaluation,"

Mathematical Software (J. Rice, ed.), New York: Academic Press, 369-390.

Bashforth, F., and Adams, J. C. (1883), *Theories of Capillary Action*, Cambridge: Cambridge University Press.

Berzins, M., and Dew, P. W. (1986), "Differential Algebraic Equations Integrators for Use with Methods of Lines," ODE Conference held at Sandia National Lab., Albuqueque, New Mexico, USA, July 1986.

Bettis, D. G. (1976), "Efficient Embedded R-K Methods," *Proceedings Numerical Treatment of DEs* (Bulirsch, Grigorieff, and Schroder, eds.), Berlin: Springer-Verlag.

Bickart, T. A. (1977), "An Efficient Solution Process for IRK Methods," *SIAM Journal on Numerical Analysis* **14**, 1022-1027.

Bickart, T. A., and Picel, Z. (1973), "High Order Stiffly Stable Composite Multistep Methods for the Numerical Integration of Stiff Differential Equations," *BIT* **13**, 272-286.

Bickart, T. A., Burgess, D. A., and Sloate, H. M. (1971), "High Order A-stable Composite Multistep Methods for the Numerical Integration of Stiff Differential Equations," *Proceedings of the Ninth Annual Allerton Conference on Circuit and System Theory*, University of Illinois at Urbana-Champaign, 465-473.

Bickart, T. A., and Jury, E. I. (1977), "Arithmetic Tests for A-stability, A(α)-stability, and Stiff-stability," *BIT* **18**, 9-21.

Bjorck, A., and Zlatev, Z. (1986), "Exploiting the Separability in the Solution of Systems of Ordinary Differential Equations," ODE Conference held at Sandia National Lab., Albuquerque, New Mexico, USA, July 1986.

Blum, E. K. (1962), "A Modification of the R-K Fourth-order Method," *Mathematics of Computation* **16**, 176-187.

Bo, H-L., Yans, W-H., and Xin, A-Y. (1983), "An Effective Method for a System of ODEs with RHS Functions Containing Discontinuities," *Proceedings*

Tenth IMACS World Congress on System Simulation and Scientific Computation, 6-9.

Bohmer, K., and Stetter, Hans J. (1984), Defect Correction Methods Theory and Applications, New York: Springer-Verlag.

Bokhoven, Van (1980), "Efficient Higher Order Implicit One-step Methods for Integration of Stiff Differential Equations," *BIT* **20**, 34-43.

Bond, J. E., and Cash, J. R. (1979), "A Block Method for the Numerical Integration of Stiff Systems of Ordinary Differential Equations," *BIT* **19**, 329-447.

Brayton, R. K., Gustavson, F. G., and Hachtel, G. D. (1972), "A New Efficient Algorithm for Solving Differential-Algebraic Systems Using Implicit BDF," *Proceedings of the IEEE* **60**, 98-108.

Brenan, K. E. (1986), "The Numerical Solution of Higher Index DAE by IRK Methods," ODE Conference held at Sandia National Lab., Albuquerque, New Mexico, USA, July 1986.

Brown, R. L. (1974), "Multiderivative Numerical Methods for Solution of Stiff ODEs," *Report No.* **UIUCDCS-R-74-672**, Department of Computer Science, University of Illinois at Urbana-Champaign, Urbana, Illinois, USA.

Brown, R. L. (1977), "Some Characteristics of Implicit Multistep Multiderivative Integration Formulas," *SIAM Journal on Numerical Analysis* **14**, 982-993.

Brown, P. N., and Hindmarsh, A. C. (1986), "Matrix-Free Methods for Stiff IVPs," ODE Conference held at Sandia National Lab., Albuquerque, New Mexico, USA, July 1986.

Brown, R. R., Riley, J. D., and Bennett, M. M. (1965), "Stability Properties of Adams-Moulton Type Methods," *Mathematics of Computation* **19**, 90-96.

Brunner, H. (1972), "A Class of A-stable Two-step Methods Based on Schur Polynomial," *BIT* **12**, 468-474.

Brusa, L., and Nigro, L. (1980), "A One-Step Method for Direct Integration of Structural Dynamic Equations," *International Journal on Numerical Methods in Engineering* **15**, 685-699.

Bui, T. D. (1979), "Some A-stable and L-stable Methods for the Numerical Integration of Stiff ODEs," *Journal of the ACM* **26**, 483-493.

Bui, T. D., and Poon, S.W.H. (1981), "On the Computational Aspect of Rosenbrock Procedures with Built-in Error Estimates for Stiff Systems," *BIT* **21**, 168-174.

Bulirsch, R., and Stoer, J. (1966a), "Numerical Treatment of ODEs by Extrapolation Methods," *Numerische Mathematik* **8**, 1-13.

Bulirsch, R., and Stoer, J. (1966b), "Asymptotic Upper and Lower Bounds for Results of Extrapolation Methods," *Numerische Mathematik* **8**, 93-104.

Burrage, K. (1978a), "A Special Family of Runge-Kutta Methods," *BIT* **18**, 22-41.

Burrage, K. (1978b), "High Order Algebraically Stable R-K Methods," *BIT* **18**, 373-383.

Burrage, K., and Butcher, J. C. (1979), "Stability Criteria for Implicit R-K Methods," *SIAM Journal on Numerical Analysis* **16**, 46-57.

Burrage, K. (1980), "Nonlinear Stability of Multivalue Multiderivative Methods," *BIT* **20**, 316-325.

Burrage, K. (1981), "Iteration Schemes for Singly-implicit R-K Methods," *Report No.* **24**, Department of Computer Science, University of Auckland, New Zealand.

Burrage, K. (1982), "Efficiently Implementable Algebraically Stable R-K Methods," *SIAM Journal on Numerical Analysis* **19**, 245-258.

Burrage, K. (1986), "The Applicability of RK Methods and Their Generalizations for Solving Stiff Ordinary Differential Equations," ODE Conference held at Sandia National Lab., Albuquerque, New Mexico, USA, July 1986.

Burrage, K. S., and Butcher, J. C. (1980), "Nonlinear Stability of a General Class of Differential Equation Methods," *BIT* **20**, 185-203.

Burrage, K., Butcher, J. C., and Chipman, F. H. (1980), "An Implementation of Singly Implicit R-K Methods, *BIT* **20**, 326-340.

Butcher, J. C. (1963), "Coefficients for the Study of R-K Integration Processes," *Journal of the Australian Mathematical Society* **3**, 185-201.

Butcher, J. C. (1964), "Implicit Runge Kutta Processes," *Mathematics of Computation* **18**, 50-64.

Butcher, J. C. (1965a), "A Modified Multistep Method for the Numerical Integration of Ordinary Differential Equations," *Journal of the ACM* **12**, 124-135.

Butcher, J. C. (1965b), "On the Attainable Order of R-K Methods," *Mathematics of Computation* **19**, 408-417.

Butcher, J. C. (1966), "On the Convergence of Numerical Solution to ODEs," *Mathematics of Computation* **20**, 1-10.

Butcher, J. C. (1967), "A Multistep Generalization of R-K Methods with Four or Five Stages," *Journal of the ACM* **14**, 84-99.

Butcher, J. C. (1972), "An Algebraic Theory of Integration Methods," *Mathematics of Computation* **26**, 79-106.

Butcher, J. C. (1973), "The Order of Numerical Methods for ODEs," *Mathematics of Computation* **27**, 793-806.

Butcher, J. C. (1975), "*A*-stability Property of Implicit R-K Methods," *BIT* **15**, 358-361.

Butcher, J. C. (1976a), "On the Implementation of Implicit R-K Methods," *BIT* **16**, 237-240.

Butcher, J.C. (1976b), "Implicit RK and Related Methods," *Modern Numerical Methods for Ordinary Differential Equations* (G. Hall and J. M. Watts, eds.), Oxford: Oxford University Press, 136-151.

Butcher, J. C. (1979), "A Transformed Implicit R-K Method," *Journal of the ACM* **2**, 731-738.

Butcher, J. C., Burrage, K., and Chipman, F. H. (1979), "STRIDE: Stable R-K Integrator for DEs," *Computational Mathematics Report No.* **20**, University of Auckland, New Zealand.

Butcher, J. C. (1981a), "Stability Properties for a General Class of Methods for ODEs," *SIAM Journal on Numerical Analysis* **18**, 37-44.

Butcher, J. C. (1981b), "AN-stability is not Equivalent to Algebraic Stability," *Computer Sciences Report No.* **26**, University of Auckland, New Zealand.

Butcher, J. C. (1982a), "A Short Proof Concerning B-stability," *BIT* **22**, 528-529.

Butcher, J. C. (1982b), "Linear and Nonlinear Stability for General Linear Methods," *Computer Science Report No.* 28, University of Auckland, New Zealand.

Butcher, J. C. (1984), "An Application of the RK Space," *BIT* **24**, 425-440.

Byrne, G. D., and Hindmarsh, A. C. (1975a), "A Poly-algorithm for Numerical Solution of Ordinary Differential Equations," *ACM Transactions on Mathematical Software* **1**, 71-96.

Byrne, G. D., and Hindmarsh, A. C. (1975b), "EPISODE: An Experimental Package for the Integration of Systems of Ordinary Differential Equations," *Report UCID-30112,* Lawrence Livermore Lab., University of California, Livermore.

Byrne, G. D., Hindmarsh, A. C., Jackson, K. R., and Brown, H. G. (1977), "A Comparison of Two ODE Solvers: GEAR and EPISODE," *Computers and Chemical Engineering* **1**, 133-147.

Byrne, G. D. (1985), "Experience at Exxon," *Proceedings of the International Conference on Stiff Computation* (R. Aiken, ed.), (April 1982; Park City, UT), 63-66, New York: Oxford Press.

Byrne, G. D., and Hindmarsh, A. C. (1985), "Experiments in Numerical Methods for a Problem in Combustion Modeling, *Journal of Applied Numerical Mathematics* **1**, 29-57.

Byrne, G. D., DeGregoria, A. J., and Salane, D. E. (1984), "A Program for Rate Constants in Gas Phase Chemical Kinetics Models," *SIAM Journal on Scientific and Statistical Computing* **5**, 642-657.

Calahan, D. A. (1967), "Numerical Solution of Linear Systems with Widely Separated Time Constants," *Proceedings of the IEEE* **5**, 2016-2017.

Campbell, S. L., and Petzold, L. R. (1982), "Canonical Forms and Solvable Systems of Differential Equations," *Report No. SAND 82-8891*, Sandia National Lab., Albuquerque, New Mexico, USA.

Carver, M. B. (1977), "Efficient Handling of Discontinuities and Time Delays in ODEs," *Proceedings of Simulation '77* (M. Hamza, ed.), Montreux, Switzerland: Acta Press.

Carver, M. B. (1978), "Efficient Integration over Discontinuities in ODE Simulations," *Mathematics of Computation Simulation 20*, **20**, 190-196.

Carver, M. B. (1979), "FORSIM IV: A Program Package for the Automatic Solution of Arbitrarily Defined Differential Equations," *Computer Physics Communication* **17**, 239-282.

Carver, M. B., and MacEwen, S. R. (1981), "On the Use of Sparse Matrix Approximation to the Jacobian in Integrating Large Sets of ODEs," *SIAM Journal on Scientific and Statistical Computing* **2**, 51-64.

Cash, J. R. (1975), "A Class of Implicit Runge-Kutta Methods for the Numerical Integration of Stiff ODEs," *Journal of the ACM* **22**, 504-511.

Cash, J. R. (1976), "Semi-implicit R-K Procedures with Error Estimates for the Numerical Solution of Stiff Systems of ODEs," *Journal of the ACM* **23**, 455-460.

Cash, J. R. (1977a), "A Class of Iterative Algorithms for the Integration of Stiff Systems of ODEs," *Journal of the Institute of Mathematics and Its Applications* **19**, 325-335.

Cash, J. R. (1977b), "On a Class of Cyclic Methods for the Numerical Integration of Stiff Systems of ODEs," *BIT* **17**, 270-280.

Cash, J. R. (1977c) "On a Class of Implicit RK Procedures," *Journal of the Institute of Mathematics and Its Applications* **19**, 455-470.

Cash, J. R. (1977d), "A Note on the Computational Aspects of a Class of IRK Procedures," *Journal of the Institute of Mathematics and Its Applications* **20**, 425-441.

Cash, J. R. (1978), "High Order Methods for the Numerical Integration of ODEs," *Numerische Mathematik* **30**, 385-409.

Cash, J. R. (1979), *Stable Recursions with Applications to the Numerical Solution of Stiff Systems,* London: Academic Press.

Cash, J. R. (1980a), "On the Integration of Stiff Systems of ODEs using Extended Backward Differentiation Formulae," *Numerische Mathematik* **34**, 235-246.

Cash, J. R. (1980b), "On the Integration of Stiff Systems of Ordinary Differential Equations Using Extended BDF," *Numerische Mathematik* **34**, 235-246.

Cash, J. R. (1981a), "Second Derivative Extended Backward Differentiation Formulas for the Numerical Integration of Stiff Systems," *SIAM Journal on Numerical Analysis* **18**, 21-36.

Cash, J. R. (1981b), "High Order P-stable Formulae for the Numerical Integration of Periodic Initial Value Problems," *Numerische Mathematik* **37**, 355-370.

Cash, J. R. (1981c), "On the Exponential Fitting of Composite Multiderivative Linear Multistep Methods," *SIAM Journal on Numerical Analysis* **18**, 808-821.

Cash, J. R. (1981d), "A Note on the Exponential Fitting of Blended, Extended LMM," *BIT* **21**, 450-454.

Cash, J. R. (1983a), "Block RK Methods for the Numerical Integration of IVPs in ODEs: The Nonstiff Case," *Mathematics of Computation* **40**, 175-191.

Cash, J. R. (1983b) "Block RK Methods for the Numerical Integration of IVPs in ODEs: The Stiff Case," *Mathematics of Computation* **40**, 193-206.

Cash, J. R. (1983c), "The Integration of Stiff IVPs in Ordinary Differential Equations Using Modified Extended BDF," *Computers and Mathematics with Applications* **9**, 645-657.

Cash, J. R.(1986), "A Block 5(4) RKF for Nonstiff IVPs," ODE Conference held at Sandia National Lab., Albuquerque, New Mexico, USA, July 1986.

Cash, J. R., and Liem, C. B. (1980), "On the Design of a Variable Step DIRK Algorithms," *Journal of the Institute of Mathematics and Its Applications* **26**, 87-91.

Cash, J. R., and Singhal, A. (1982), "Mono-implicit RKF for the Numerical Integration of Stiff Differential Equations," *IMA Journal on Numerical Analysis* **2**, 211-217. Cellier, F. E., and Rufer, D. F. (1978), "Algorithm Suited for the Solution of IVPs in Engineering Application," *Math. Comput. Simulation* **XX**, 160-165.

Cellier, F. E. (1983), "Stiff Computation, Where to Go," *Proceedings of the International Conference on Stiff Computation* (R. Aiken, ed.), (April 1982; Park City, UT), New York: Oxford.

Ceschino, F., and Kuntzmann, J. (1966), *Numerical Solution of IVPs,* Englewood Clifs, New Jersey: Prentice-Hall.

Chang, Y. F. (1974), "Automatic Solution of Differential Equations," *Lecture Notes in Mathematics* (D. L. Colton and R. P. Gilbert, eds.) **430**, New York: Springer-Verlag, 61-94.

Chang, Y. F., and Corliss, G. F. (1980), "Ratio-like and Recurrence Relation Tests for Convergence of Series," *Journal of the Institute of Mathematics and Its Applications* **25**, 349-359.

Chang, Y. F., Corliss, G. F., and Morris, R. (1979), "ATSMCC User Manual," Department of Mathematics, Statistics and Computer Science Techical Report, Marquette University, Wisconsin.

Chang, Y. F. (1986), "Solving Stiff Systems by Taylor Series," ODE Conference held at Sandia National Lab., Albuquerque, New Mexico, USA, July 1986.

Chase, P. E. (1962), "Stability Properties of Predictor-Corrector Methods for ODEs," *Journal of the ACM* **9**, 457-468.

Chawla, M. M. (1981), "Two-step Fourth Order P-stable Methods for Second Order Differential Equations," *BIT* **21**, 190-193.

Chawla, M. M. (1983), "Unconditional Stability of Numerov-type Methods for Second Order Differential Equations," *BIT* **23**, 541-542.

Chawla, M. M. (1984), "Numerov Made Explicit has Better Stability," *BIT* **24**, 117-118.

Chawla, M. M. (1985a), "On the Order and Attainable Intervals of Periodicity of Explicit Nystrom Methods for $y'' = f(t, y)$," *SIAM Journal on Numerical Analysis* **22**, 127-131.

Chawla, M. M. (1985b), "A New Class of Explicit Methods for $y'' = f(t, y)$ with Extended Intervals of Periodicity," *Journal of Computational and Applied Mathematics* **14**

Chawla, M. M., and Sharma, S. R. (1981a), "Intervals of Periodicity and Absolute Stability of Explicit Nystrom Methods for $y'' = f(x, y)$," *BIT* **21**, 455-464.

Chawla, M. M., and Sharma, S. R. (1981b), "Families of Fifth Order Nystrom Methods for $y'' = f(x, y)$ and Intervals of Periodicity," *Computing* **26**, 247-256.

Chawla, M. M., and Sharma, S. R. (1985), "Families of Three-Stage Third Order Runge-Kutta-Nystrom Methods for $y'' = f(x, y, y')$," *Journal of the Australian Mathematical Society* **26**, 375-386.

Chawla, M. M., and Subramanian, R. (1984), "Regions of Absolute Stability of Explicit Runge-Kutta-Nystrom Methods for $y'' = f(x, y, y')$," *Journal of Computational and Applied Mathematics* **11**, 259-266.

Chawla, M. M., and Rao, P. S. (1985), "High Accuracy P-stable Methods for $y'' = f(x,y)$," *IMA Journal on Numerical Analysis* **5**, 215-220.

Chawla, M. M., and Rao, P. S. (1986), "Phaselag Analysis of Explicit Nystrom Methods for $y'' = f(x,y)$," *BIT* **26**, 64-70.

Chawla, M. M., and McKee, S. (1986), "Convergence of Linear Multistep Methods with Multiple Roots," *Journal of Computational and Applied Mathematics* **14**, 451-454.

Chawla, M. M., and Neta, B. (1986), "Families of Two-Step Fourth Order P-Stable Methods for Second Order Differential Equations," *Journal of Computational and Applied Mathematics*

Chawla, M. M., and Sharma, S. R. (1980), "Families of Direct Fourth-Order Methods for the Numerical Solution of General Second Order IVPs," *ZAMMM* **60**, 469-488.

Chipman, F. H. (1971), "A-stable Runge-Kutta Processes," *BIT* **11**, 384-388.

Chipman, F. H. (1973), "The Implementation of R-K Implicit Processes," *BIT* **13**, 391-393.

Chipman, F. (1986), "Stability Properties of Several Classes of General Linear Methods," ODE Conference held at Lab., Albuquerque, New Mexico, USA, July 1986.

Chu, M. T., and Hamilton, H. (1986), "Parallel Processing for the Numerical Solution of ODE," ODE Conference held at Sandia National Lab., Albuquerque, New Mexico, USA, July 1986.

Chu, M. T., and H. Hamilton (1987), "Parallel Solution of Ordinary Differential Equations by Multiblock Methods," *SIAM Journal on Scientific and Statistical Computing* **8**, 342-353.

Clark, K. (1986) "Canonical Forms for Differential Algebraic Systems Solvable by the BDF," ODE Conference held at Sandia National Lab., Albuquerque, New Mexico, USA, July 1986.

Coddington, E. A., and Levinson, N. (1955), *Theory of Ordinary Differential Equations,* New York: McGraw-Hill.

Consdine, S. (1986), "On the Integration of Stiff Differential Systems Using Modified Extended BDF," ODE Conference held at Sandia National Lab., Albuquerque, New Mexico, USA, July 1986.

Conte, S. D., and Reeves, R. F. (1956), "A Kutta Third-order Procedure for Solving DE Requiring Minimum Storage," *Journal of the ACM* **3**, 22-25.

Cooper, G. J., and Sayfy, A. (1979), "Semiexplicit A-stable R-K Methods," *Mathematics of Computation* **33**, 541-556.

Cooper, G. J. (1981), "Error Estimates for General Linear Methods for ODEs," *SIAM Journal on Numerical Analysis* **18**, 65-82.

Cooper, G. J., and Butcher, J. C. (1983), "An Iteration Scheme for Implicit R-K Methods," *IMA Journal on Numerical Analysis* **3**, 127-140.

Corliss, G. F., and Lowery, D. (1977), "Choosing a Stepsize for Taylor Series Methods for Solving Ordinary Differential Equations" *Journal of Computational and Applied Mathematics* **3**, 251-256.

Corliss, G. F., and Chang, Y. F. (1982), "Solving Ordinary Differential Equations Using Taylor Series," *ACM Transactions on Mathematical Software* , 114-144.

Crane, R. C., and Klopfenstein, R. W. (1965), "A Predictor-Corrector Algorithm with Increased Range of Absolute Stability," *Journal of the ACM* **12**, 227-241.

Creedon, D. M., and Miller, J. J. (1975), "The Stability Properties of q-step Backward Difference Schemes," *BIT* **15**, 244-249.

Crouzeix, M., and Lisbona, F. J. (1984), "The Convergence of Variable Stepsize, Variable Formula, Multistep Methods," *SIAM Journal on Numerical Analysis* **21**, 512-534.

Cryer, C. W. (1972), "Numerical Methods for Functional Differential Equations," *Proceedings International Conference on Stiff Computation,* New York: Academic Press, 17-102.

Cryer, C. W. (1973), "A New Class of Highly Stable Methods: A_0-stable Methods," *BIT* **13**, 153-159.

Curtis, C. F., and Hirschfelder, J. O. (1952), "Integration of Stiff Equations," *National Academy of Sciences* **38**, 235-243.

Curtis, A. R., and Reid, J. K. (1974), "The Choice of Steplength when using Differences to Approximate Jacobian Matrices," *Journal of the Institute of Mathematics and Its Applications* **13**, 121-126.

Curtis, A. R. (1980), "FACSIMILE: Numerical Integrator for Stiff IVPs," *Conference on Computational Techniques for ODEs* (I. Gladwell and D. K. Sayers, eds.), New York: Academic Press, 47-82.

Dahlquist, G. (1956), "Convergence and Stability in the Numerical Integration of ODEs," *Math. Scand.* **4**, 33-53.

Dahlquist, G. (1959), "Stability and Error Bounds in the Numerical Integration of ODEs," *Transcript* **130**, Royal Institute of Technology, Stockholm.

Dahlquist, G. (1963), "A Special Stability Problem for Linear Multistep Methods," *BIT* **3**, 27-43.

Dahlquist, G. (1969), "A Numerical Method for Some ODEs with Large Lipschitz Constants," *Information Processing* **68**, Amsterdam: North-Holland Publishing Co.

Dahlquist, G. (1974), "Problems Related to the Numerical Treatment of Stiff DEs," *International Computing Symposium* (A. Gunter, ed.), Amsterdam: North-Holland Publishing Co., 307-314.

Dahlquist, G., and Bjorck, A. (1974), *Numerical Methods,* Englewood Cliffs, New Jersey: Prentice-Hall.

Dahlquist, G. (1975a), "Error Analysis for a Class of Methods for Stiff IVPs," Dundee Conference on Numerical Analysis, *Lecture Notes in Mathematics* **506**, Berlin: Springer-Verlag, 60-74.

Dahlquist, G. (1975b), "On Stability and Error Analysis for Stiff Nonlinear Problems," *Report TRITA-NA-7508,* Department of Information Processing and Computer Science, Royal Institute of Technology, Stockholm.

Dahlquist, G. (1975c), "Recent Work on Stiff DEs," *Report TRITA-NA-7512,* Department of Information Processing, Royal Institute of Technology, Stockholm.

Dahlquist, G. (1976), "On the Relation of *G*-stability to other Stability Concepts for LMM," *Report TRITA-NA-7621,* Department of Information Processing, Royal Institute of Technology, Stockholm.

Dahlquist, G. (1978a), "*G*-stability is Equivalent to *A*-stability," *BIT* **18**, 384-401.

Dahlquist, G. (1978b), "On Accuracy and Unconditional Stability of LMM for Second Order Differential Equations," *BIT* **18**, 133-136.

Dahlquist, G. (1979a), "Some Properties of Linear Multistep and One-leg Methods for ODEs," *Report TRITA-NA-7904,* Department of Computer Science, Royal Institute of Technology, Stockholm.

Dahlquist, G. (1979b), "Some Contractivity Questions for One-leg and Linear Multistep Methods," *Report TRITA-NA-7905,* Royal Institute of Technology, Stockholm.

Dahlquist, G., Edsberg, L. Skollermo, G., and Soderlind, G. (1980), "Are the Numerical Methods and Software Satisfactory for Chemical Kinetics?," *Report TRITA-NA-8005,* Royal Institute of Technology, Stockholm; also, in Proceedings of the Workshop on Numerical Integration of Differential Equations, Universitat Bielefeld, Zentrum fur Interdis Ziphnare, Forschung (April 14-23).

Dahlquist, G. (1981), "On the Local and Global Errors of One-leg Methods," *Report TRITA-NA-8110,* Royal Institute of Technology, Stockholm.

Dahlquist, G. (1983a), "On One-leg Multistep Methods," *SIAM Journal on Numerical Analysis* **20**, 1130-1146.

Dahlquist, G. (1983b), "On Matrix Majorants and Minorants, with Applications to Differential Equations," *Linear Algebra and Its Applications* **52/53**, 199-216.

Dahlquist, G., and Soderlind, G. (1982), "Some Problems Related to Stiff Nonlinear Differential Systems" *Computational Methods in Applied Science and Engineering* **V** (R. Glowinski and J. L. Lions, eds.), Amsterdam: North-Holland Publishing Co.

Dahlquist, G., Liniger, W., and Nevanlinna, O. (1983), "Stability of Two-step Methods for Variable Integration Steps," *SIAM Journal on Numerical Analysis* **20**, 1071-1085.

Dahlquist, G., Mingyou, H., and LeVeque, R. (1983). "On the Uniform Power-Boundedness of a Family of Matrices and the Applications to One-Leg and Linear Multistep Methods," *Numerische Mathematik* **42**, 1-13.

Dalle, R. L., and Pasciutti, F. (1975), "Runge-Kutta Methods with Global Error Estimates," *Journal of the Institute of Mathematics and Its Applications* **16**, 381-388.

Daniel, J. W., and Moore, R. E. (1970), *Computation and Theory of Ordinary Differential Equations,* San Francisco: W. H. Freeman and Co.

Dekker, K., and Verwer, J. G. (1984), *Stability of Runge-Kutta Methods for Stiff Nonlinear Differential Equations,* Amsterdam: North-Holland Publishing Co.

Deuflhard, P. (1983), "Order and Stepsize Control in Extrapolation Methods," *Numerische Mathematik* **41**, 399-422.

Deuflhard, P. (1984), "Computation of Periodic Solutions of Nonlinear ODEs," *BIT* **24**, 456-466.

Deuflhard, P. (1985), "Recent Progress in Extrapolation Methods for ODEs," *SIAM Review* **27**, 505-536.

Deuflhard, P., Bader, G., and Nowak, U. (1980), "LARKIN--A Software Package for the Numerical Simulation of LARge Systems Arising in Chemical Reaction KINetics," *Report No.* **100**, University of Heidelberg, W. Germany.

Deuflhard, P., and Bader, G. (1982), "Multiple Shooting Techniques Revisited," *Report No.* **163**, Institut fur Angewandte Mathematics, University of Heidelberg, W. Germany.

Doolan, E. P., Miller, J.J.H., and Schilder, W.H.F. (1980), *Uniform Numerical Methods for Problems with Initial and Boundary Layers,* Dublin: Boole Press.

Donelson, J., and Hansen, E. (1971), "Cyclic Composite Multistep Methods," *SIAM Journal on Numerical Analysis* **8**, 137-157.

Dongarra, J. J., Bunch, J. R., Moler, C. B., and Stewart, G. W. (1979), "LINPACK User's Guide," *SIAM: Philadelphia.*

Dormand, J. R., and Prince, P. J. (1980), "A Family of Embedded RKF's," *Journal of Computational and Applied Mathematics* **6**, 19-26.

Dormand, J. R., Duckers, R. R., and Prince, P. J. (1984), "Global Error Estimation With RK Methods," *SIAM Journal on Numerical Analysis* **42**, 169-184.

Dormand, J. R., and Prince, P. J. (1985), "Global Error Estimation With RK Methods II," *IMA Journal on Numerical Analysis* **5**, 481-497.

Dormand, J. R., and Prince, P. J. (1987), "Runge-Kutta-Nystrom Triples," *Computers and Mathematics with Applications* **13**, 937-949.

Duff, I. F., Erisman, A. M., Gear, C. W., and Reid, J. K. (1985), "Some Remarks on the Inverse of Sparse Matrices," Mathematics and Computer Science Division Report No. **51**, Argonne National Lab., Argonne, Illinois, USA.

Ehle, B. L. (1968), "High Order A-stable Methods for the Numerical Solution of Differential Equations," *BIT* **8**, 276-278.

Ehle, B. L. (1973), "A-stable Methods and PadéApproximations to the Exponential," *SIAM Journal on Mathematical Analysis* **4**, 671-680.

Ehle, B. L., and Picel, Z. (1975), "Two Parameter, Arbitrary Order, Exponential Approximation for Stiff Equations," *Mathematics of Computation* **29**, 501-511.

Eisenstat, S. C., Gursky, V. C., Schultz, M., and Sherman, A. H. (1977a), "Yale Sparse Matrix Package I: The Symmetric Codes," *Research Report No.* **112**, Department of Computer Science, Yale University.

Eisenstat, S. C., Gursky, V. C., Schultz, M., and Sherman, A. H. (1977b), "Yale Sparse Matrix Package II: The Nonsymmetric Codes," *Research Report No.* **114**, Department of Computer Science, Yale University.

Ellison, D. (1981), "Efficient Automatic Integration of ODEs with Discontinuities," *Mathematics of Computation Sim. 23,* **23**, 12-20.

England, R. (1969), "Error Estimates for R-K Type Solutions to Systems of ODEs," *Computer Journal* **12**, 166-175.

Enright, W. H. (1972), "Studies in the Numerical Solution of Stiff ODEs," Ph.D. thesis, *Report No.* **46**, Department of Computer Science, University of Toronto, Canada.

Enright, W. H. (1974a), "Second Derivative Multistep Methods for Stiff ODEs," *SIAM Journal on Numerical Analysis* **11**, 321-331.

Enright, W. H. (1974b), "Optimal Second Derivative Methods for Stiff Systems," *Conference on Stiff Differential Systems* (R. A. Willoughby, ed.), New York: Plenum Publishing co., 95-109.

Enright, W. H., Hull, T. E., and Lindberg, B. (1975), "Comparing Numerical Methods for Stiff Systems of ODEs," *BIT* **15**, 10-48.

Enright, W. H., and Hull, T. E. (1976), "Test Results on Initial Value Methods for Non-stiff ODEs," *SIAM Journal on Numerical Analysis* **13**, 944-961.

Enright, W. H. (1978), "Improving the Efficiency of Matrix Operations in the Numerical Solution of Stiff ODEs," *ACM Transactions on Mathematical Software* **4**, 127-136.

Enright, W. H. (1979), "Using a Testing Package for the Automatic Assessment of Numerical Methods for ODEs," *Performance Evaluation of Numerical Methods* (L. D. Fosdick, ed.), Amsterdam: North-Holland Publishing Co., 199-213.

Enright, W. H., and Kamel, M. S. (1979), "Automatic Partitioning of Stiff Systems and Exploiting the Resulting Structure," *ACM Transactions on Mathematical Software* **5**, 374-385.

Enright, W. H., and Kamel, M. S. (1980), "On Selecting a Low Order Model using the Dominant Mode Concept," *IEEE Trans. Automatic Contr.* **AC-25** 5, 976-978.

Enright, W. H. (1983a), Private communication.

Enright, W. H. (1983b), "Exploiting Initial Value Expertise in the Solution of 2-point BVP," *Proceedings of Computational Mathematics I* (S. O. Fatunla, ed.), 11-18, Dublin: Boole Press.

Enright, W. H., and Pryce, J. D. (1983), "Two FORTRAN Packages for Assessing IV Methods," *Technical Report No.* **167/83**, Department of Computer Science, University of Toronto, Canada.

Enright, W. H., and Addison, C. A. (1984), "Properties of Multistep Formulas Intended for a Class of Second Order ODEs," *SIAM Journal on Numerical Analysis* **20**, 327-339.

Enright, W. H. (1986), "A New Error Control for IVPs," ODE Conference held at Sandia National Lab., Albuquerque, New Mexico, USA, July 1986.

Enright, W. H., Jackson, K. R., Norsett, S. P., and Thomsen, P. G. (1986), "Effective Solution of Discontinuous IVPs Using a RKF pair with Interpolants," ODE Conference held at Sandia National Lab., Albuquerque, New Mexico, USA, July 1986.

Enright, W. H., and Pryce, J. D. (1987), "Two FORTRAN Packages for Assessing Initial Value Methods," *ACM Transactions on Mathematical Software* **13**, 1-22.

Euler, L. (1913), "Opera Omnia, Series Prima," **11**, Leipzig and Berlin.

Euler, L. (1914), "Opera Omnia, Series Prima," **12**, Leipzig and Berlin.

Evans, D. J., and Fatunla, S. O. (1975), "Accurate Numerical Determination of the Intersection Point of the Solution of a DE with a Given Algebraic Relation," *Journal of the Institute of Mathematics and Its Applications* **16**, 355-359.

Fatunla, S. O. (1976), "A New Algorithm for the Numerical Solution of ODEs," *Computers and Mathematics with Applications* **2**, 247-253.

Fatunla, S. O. (1978a), "An Implicit Two-Point Numerical Integration Formula for Linear and Nonlinear Stiff Systems of ODEs," *Mathematics of Computation* **32**, 1-11.

Fatunla, S. O. (1978b), "A Variable Order One Step Scheme for Numerical Solution of ODEs," *Computer and Mathematics with Applications* **4**, 33-41.

Fatunla, S. O. (1980), "Numerical Integrators for Stiff and Highly Oscillatory Differential Equations," *Mathematics of Computation* **34**, 373-390.

Fatunla, S. O. (1982a), "Nonlinear Multistep Methods for Initial Value Problems," *Computers and Mathematics with Applications* **8**, 231-239.

Fatunla, S. O. (1982b), "Numerical Treatment of Special IVPs," *Proceedings BAIL II Conference* (J.J.H. Miller, ed.), Dublin: Trinity College, 28-45.

Fatunla, S. O. (1984), "One Leg Multistep Method for Second Order Differential Equations," *Computers and Mathematics with Applications* **10**, 1-4.

Fatunla, S. O. (1985), "One Leg Hybrid Formula for Second Order IVPs," *Computers and Mathematics with Applications* **10**, 329-333.

Fatunla, S. O. (1985c), *Computational Mathematics I, Proceedings First International Conference on Numerical Analysis and its Applications,* (November 1983; Benin City, Nigeria), Dublin: Boole Press, 25-31.

Fatunla, S. O. (1986), "Numerical Treatment of Singular IVPs," *Computers and Mathematics with Applications* **12**, 1109-1115.

Fatunla, S. O. (1987a), "Hybrid Formulas with Nonvanishing Interval of Periodicity," *International Journal of Computer Mathematics* **21**, 145-159.

Fatunla, S. O. (1987b), "*P*-stable Methods for Second Order IVPs," *Proceedings Conference on Computational Mathematics* (S. O. Fatunla, ed.), University of Benin, Benin City, Nigeria, Dublin: Boole Press, 25-31.

Fatunla, S. O. (1987c), "New Predictor Corrector Formulas for IVPs in ODEs," Journal of Computer Mathematics **24**, to appear.

Fatunla, S. O. (1987d), *Computational Mathematics II, Proceedings Second International Conference on Numerical Analysis and its Applications,* (January 1986; Benin City, Nigeria), Dublin: Boole Press, 25-29.

Fatunla, S. O. (1987e), *Recent Advances in Stiff ODE Solvers* (S. O. Fatunla, ed.), University of Benin, Benin City, Nigeria, Dublin: Boole Press (to appear).

Fehlberg, E. (1964), "New High Order R-K Formulas with Stepsize Control for Systems of First and Second Order DE," *Z. Angew. Math. Mech.* **44**, 17-29.

Fehlberg, E. (1968), "Classical Fifth, Sixth, Seventh and Eighth Order Formulas with Stepsize Control," *NASA Technical Report No.* **287**, George C. Marshall Space Flight Center, Huntsville.

Fehlberg, E. (1970), "Wassische Runge-Kutta-Formula Funfter and Siebenter Ordnung mit Schrittweiten-Kontrolle," *Computing* **4**, 9-106.

Fehlberg, E. (1969), "Low Order Classical Runge-Kutta Formulas with Stepsize Control and their Application to Some Heat Transfer Problems," *NASA Technical Report No.* 315, George C. Marshall Space Flight Center, Huntsville.

Feinberg, R. B. (1982), "A_0-Stable Formulas of Adams Type," *SIAM Journal on Numerical Analysis* **19**, 259-262.

Finden, W. F. (1975), "Some Numerical Procedures for Solving Systems of ODEs Containing a Small Parameter," *Research Report* **CS-75-22**, Department of Computer Science, University of Waterloo, Canada.

Fine, J. (1986), "Evaluation and Implementation of RK Nystrom Methods," ODE Conference held at Sandia National Lab., Albuquerque, New Mexico, USA, July 1986.

Fletcher, R. (1980), *Practical Methods of Optimization,* Volume 1, *Unconstrained Optimization,* New York: John Wiley & Sons.

Forsythe, G. E., Malcolm, M. A., and Moler, C. B. (1977), Computer Methods for Mathematical Computation, Englewood Cliffs: Prentice-Hall.

Fox, L., and Parker, I. B. (1968), *Chebyshev Polynomials in Numerical Analysis,* Oxford: Oxford University Press.

Franklin, M. A. (1978), "Parallel Solutions of ODEs," *IEEE Transactions on Computers* **C-27**, 413-420.

Fyfe, D. J. (1966), "Economical Evaluation of R-K Formulas," *Mathematics of Computation* **20**, 392-398.

Gaffney, P. W. (1984), "A Performance Evaluation of Some FORTRAN Subroutines for the Solution of Stiff Oscillatory ODEs," *ACM Transactions on Mathematical Software* **10**, 58-72.

Gautschi, W. (1961), "Numerical Integration of ODEs Based on Trigonometric Polynomials," *Numerische Mathematik* **3**, 381-397.

Gear, C. W. (1965), "Hybrid Methods for IVPs in ODEs," *SIAM Journal on Numerical Analysis* **2**, 69-86.

Gear, C. W. (1967), "The Numerical Integration of ODEs," *Mathematics of Computation* **21**, 146-156.

Gear, C. W. (1969), "The Automatic Integration of Stiff ODEs," in *Information Processing* **68**, (A.J.H. Morrell, ed.), Amsterdam: North-Holland Publishing Co., 187-193.

Gear, C. W. (1971a), "Algorithm 407: DIFSUB for Solution of ODEs," *Communications of the ACM*, **14**, 185-190.

Gear, C. W. (1971b), *Numerical Initial Value Problems in Ordinary Differential Equations,* Englewood Cliffs, New Jersey: Prentice-Hall.

Gear, C. W. (1971c), "Experience and Problems with the Software for the Automatic Solution of Ordinary Differential Equations," *Mathematical Software* (J. Rice, ed.), New York: Academic Press, 211-227.

Gear, C. W. (1971d), "The Simultaneous Numerical Solution of DAE's," *IEEE Transactions on Circuits and Systems* **18**, 89-95.

Gear, C. W. (1974), "Asymptotic Estimation of Errors and Derivatives for Numerical Solution of ODEs," in *Information Processing* **74**, 447-451.

Gear C. W., and Tu, K. W. (1974), "The Effect of Variable Meshsize on the Stability of Multistep Methods," *SIAM Journal on Numerical Analysis* **11**, 1025-1043.

Gear, C. W., and Watanabe, D. S. (1974), "Stability and Convergence of Variable Order Multistep Methods," *SIAM Journal on Numerical Analysis* **11**, 1044-1058.

Gear, C. W. (1976), "Future Developments in Stiff Integration Techniques-- Stability of Methods that do not use an Exact Jacobian," *Report No.* **UIUCDCS-R-76-839**, Department of Computer Science, University of Illinois at Urbana-Champaign.

Gear, C. W. (1978), "The Stability of Numerical Methods for Second Order ODEs," *SIAM Journal on Numerical Analysis* 5, 188-197.

Gear, C. W. (1980a), "Method and Initial Stepsize Selection in Multistep ODE Solvers," *Report No.* **UIUCDCS-R-80-1006**, Department of Computer Science, University of Illinois at Urbana-Champaign.

Gear, C. W. (1980b), "Automatic Detection and Treatment of Oscillatory and/or Stiff Ordinary Differential Equations," *Report No.* **UIUCDCS-R-80-1019**, Department of Computer Science, University of Illinois at Urbana-Champaign.

Gear, C. W. (1980c), "Unified Modified Divided Difference Implementation of ADAMS and BDF Formulas," *Report No.* **UIUCDCS-R-80-1014**, Department of Computer Science, University of Illinois at Urbana-Champaign.

Gear, C. W. (1980d), "Runge-Kutta Starters for Multistep Methods," *ACM Transactions on Mathematical Software* 6, 263-279.

Gear, C. W. (1981), "Numerical Solution of ODEs: Is there anything left to do?," *SIAM Review* 23, 10-24.

Gear, C W. (1982), "Rootfinding in Simulation and ODE Solution," *Report No.* **UIUCDCS-R-82-1116**, Department of Computer Science, University of Illinois at Urbana-Champaign.

Gear, C. W., and Petzold, L. R. (1982), "Singular Implicit ODEs and Constraints," *Lecture Notes in Mathematics* **1005**, 120-127.

Gear, C. W. (1983), "Stiff Software: What Do We Have and What Do We Need?," *Proceedings International Conference on Stiff Computation* (R. Aiken, ed.), (April 1982, Park City, UT), New York: Oxford.

Gear, C. W., and Saad, Y. (1983), "Iterative Solution of Linear Equations in ODE Codes," *SIAM Journal on Scientific and Statistical Computing* 4, 583-601.

Gear, C. W., and Østerby, O. (1984), "Solving ODEs with Discontinuities," *ACM Transactions on Mathematical Software* 10, 23-44.

Gear, C. W., and Wells, D. R. (1984), "Multirate Linear Multistep Methods," *BIT* 24, 484-502.

Gear, C. W., and Petzold, L. R. (1984), "ODE Methods for the Solution of Differential Algebraic Systems," *SIAM Journal on Numerical Analysis* 20, 716-728.

Gear, C. W. (1987), "The Potential for Parallelism in Ordinary Differential Equations," *Proceedings of Second International Conference on Computational Mathematics and its Applications*

Gear, C. W., and Wang, D. (1986), "Real-Time Integration Formulas with Off-step Inputs and Their Stability," ODE Conference held at Sandia National Lab., Albuquereque, New Mexico, USA, July 1986. (January 1986; Benin City,

Nigeria), (S. O. Fatunla, ed.), Dublin: Boole Press.

Gear, C. W., and D. Wang (1987), ''Explicit Stiff Stability via Splitting and the Parallel Solution of Ordinary Differential Equations,'' *Report No.* **UIUCDCS-R-87-1328**, Department of Computer Science, University of Illinois at Urbana-Champaign.

Gill, S. (1951), ''A Process for the Step-by-step Integration of Differential Equations in Automatic Digital Computing Machine,'' *Proceedings Cambridge Philosophical Society* **47**, 95-108.

Gladwell, I. (1979), ''Initial Value Routines in LAG Library,'' *ACM Transactions on Mathematical Software* **5**, 386-400.

Gladwell, I., and Sayers, D. K. (1980), *Computational Techniques for Ordinary Differential Equations,* New York: Academic Press.

Gladwell, I., Branklin, R., and Shampine, L. F. (1986), ''A Robust Method for Determining all the Roots and Turning Points of the Solution of an ODE IVPs,'' ODE Conference held at Sandia National Lab., New Mexico, USA, July 1986.

Gladwell, I., Shampine, L. F., and Brankin, R. W. (1987), ''Automatic Selection of Initial Stepsize for an ODE Solver,'' *Journal on Computational and Applied Mathematics* **18**, 175-192.

Golub, G. H., and Sloan, C.F.V. (1983), *Matrix Computations,* Baltimore: Johns Hopkins University Press.

Gourlay, A. R. (1970), ''A Note on Trapezoidal Methods for the Solution of IVPs,'' *Mathematics of Computation* **24**, 629-633.

Goyal S., and Serbin, S. M. (1987), ''A Class of Rosenbrock-type Schemes for Second Order Nonlinear Systems of ODEs,'' *Computers and Mathematics with Applications* **13**, 351-362.

Graff, O. F. (1973), ''Methods of Orbit Computation with Multirevolution Steps,'' *Report* **AMRL 1063**, Applied Mechanics Research Lab., University of Texas, Austin.

Graff, O. F., and Bettis, D. G. (1975), ''Modified Multirevolution Integration Methods for Satellite Orbit Calculations,'' *Celestial Mechanics* **11**, 443-448.

Gragg, W. B. (1964), *Repeated Extrapolation to the Limit in The Numerical Solution of Ordinary Differential Equations,* Ph.D Dissertation, University of California, Los Angeles.

Gragg, W. B. (1965), ''On Extrapolation Algorithms for Ordinary IVPs,'' *SIAM Journal on Numerical Analysis* **2**, 384-403.

Gragg, W., and Stetter, H. J. (1964), ''Generalized Multistep Predictor-Corrector Methods,'' *Journal of the ACM* **11**, 188-209.

Grigorieff, R. D. (1983), ''Stability of Multistep Methods on Variable Grids,'' *Numerische Mathematik* **42**, 359-377.

Grigorieff, R. D., and Paes-Leme, P. J. (1984), "On the Zero-Stability of the 3-Step BDF on Nonuniform Grids," *BIT* **24**, 85-91.

Gupta, G. K. (1976), "Some New High Order Multistep Formulae for Solving Stiff Equations," *Mathematics of Computation* **30**, 417-432.

Gupta, G. K. (1978), "Implementing Second Derivative Multistep Predictor Corrector Methods," *Journal of the ACM* **11**, 188-209.

Gupta, G. K. (1979), "A Polynomial Representation of Hybrid Methods for Solving ODEs," *Mathematics of Computation* **33**, 1251-1256.

Gupta, G. K. (1985), "Performance Evaluation of the Stiff ODE Code DSTIFF," *SIAM Journal on Scientific and Statistical Computing* **6**, 939-950.

Gupta, G. K., Sacks-Davis, R., and Tischer, T. (1982), "A Review of Recent Development in Solving ODEs," *Technical Report No.* **24**, Department of Computer Science, Monash University, Australia

Gupta, G. K. (1983), "Description and Evaluation of a Stiff ODE Code DSTIFF," Proceedings of Conference on Stiff Computation (R. Aiken, ed.), (April 1982; Park City, UT), New York: Oxford.

Gupta, G. K., Gear, C. W., and Leimkuhler, B. (1985), "Implementing Linear Multistep Formulas for Solving DAE's," *Report No.* **UIUCDS-R-85-1205**, Department of Computer Science, University of Illinois at Urbana-Champaign.

Haines, C. F. (1968), "Implicit Integration Processes with Error Estimates for the Numerical Solution of DEs," *Computer Journal* **12**, 183-187.

Hairer, E., and Wanner, G. (1976), "A Theory for Nystrom Methods," *Numerische Mathematik* **25**, 383-400.

Hairer, E. (1977), "Methods de Nystrom pour l'equation Differentielle $y'' = f(x, y)$," *Numerische Mathematik* **27**, 283-300.

Hairer, E. (1978), "A Runge-Kutta Method of Order 10," *Journal of the Institute of Mathematics and Its Applications* **21**, 47-59.

Hairer, E. (1979), "Unconditionally Stable Methods for Second Order DEs," *Numerische Mathematik* **32**, 373-379.

Hairer, E., and Wanner, G. (1981), "Algebraically Stable and Implementable R-K Methods of High Order," *SIAM Journal on Numerical Analysis* **18**, 1098-1108.

Hairer, E., Bader, G., and Lubich H. (1982), "On the Stability of Semi-implicit Methods for ODEs," *BIT* **22**, 211-232.

Hairer, E. (1983), "On the Instability of BDF," *SIAM Journal on Numerical Analysis* **20**, 1206-1209.

Hairer, E., and Lubich, C. (1984), "Asymptotic Expansions of the Global Error of Fixed-stepsize Methods," *Numerische Mathematik* **45**, 345-360.

Halin, H. J. (1976), "Integration of ODEs Containing Discontinuities," *Proceedings Summer Computer Science Conference,* La Jolla, California: SCI Press.

Halin, H. J. (1983), "The Application of Taylor Series Methods in Simulation," *Proceedings of Summer Computer Simulation Conference II*, Vancouver, B.C., Canada, July 11-13, Society for Computer Simulation, 1032-1076.

Hall, G. (1967), "The Stability of Predictor-Corrector Methods," *Computer Journal* 9, 410-412.

Hall, G., and Watt, J. M. (1976), *Modern Numerical Methods for ODEs*, Oxford: Clarendon Press.

Hall, G. (1985), "Equilibrium States of Runge-Kutta Schemes," *ACM Transactions on Mathematical Software* 11, 289-301.

Hall, G. (1986) "Equilibrium States of Runge-Kutta Schemes: Part II," *ACM Transacations on Mathematical Software* 12, 183-192.

Hall, G. (1986c), "Properties of Stability Regions for RK Codes," ODE Conference held at Sandia National Lab, Albuquerque, USA, July, 1986.

Hall, G., and Suleiman, M. B. (1981), "Stability of Adams-type Formulae for Second Order ODEs," *IMA Journal on Numerical Mathematics* 1, 427-438.

Hammer, P. C., and J. W. Hollingsworth (1955), "Trapezoidal Methods of Approximating Solutions of Differential Equations," *MTAC* 9, 92-96.

Hanson, P. M., and Enright, W. H. (1983), "Controlling the Defect in Existing Variable-Order Adams Codes for Initial-Value Problems," *ACM Transactions on Mathematical Software* 9, 71-97.

Hay, J. L., Gosbie, R. E., and Chaplin, R. I. (1974), "Integration Routines for Systems with Discontinuities," *Computer Journal* 17, 275-278.

Henrici, P. (1962), *Discrete Variable Methods in ODEs*, New York: John Wiley & Sons.

Hindmarsh, A. C. (1974), "GEAR: ODE System Solver," Revision 3, *Report No.* **UCID-30001**, Lawrence Livermore Lab., University of California, Livermore.

Hindmarsh, A. C. (1979), "A User Interface Standard for ODE Solvers," *ACM SIGNUM Newsletter* **14:2**, 11.

Hindmarsh, A. C. (1980), "LSODE and LSODI: Two New Initial Value ODE Solvers," *ACM SIGNUM Newsletter* 15, 10-11.

Hindmarsh, A. C. (1981), "ODE Solvers for Use with the Method of Lines," in *Advances in Computer Methods for PDE IV* (R. Vichnevetsky and R. S. Stepleman, eds.), New Brunswick: IMACS.

Hindmarsh, A. C. (1983a), "Toward a Systematized Collection of ODE Solvers," *World Congress on Systems Simulation and Scientific Computation*, Amsterdam: North-Holland, 427-429.

Hindmarsh, A. C. (1983b), "Stiff System Problems and Solutions at Lawrence Livermore National Lab.," *Proceedings of International Conference on Stiff Computation* (R. Aiken, ed.), (April 1982; Park City, UT), New York: Oxford University Press.

Jackson, K. R., and Sacks-Davis, R. (1980), "An Alternative Implementation of Variable Stepsize Multistep Formulas for Stiff ODEs," *ACM Transactions on Mathematical Software* **6**, 295-318.

Jain, M. K., Jain, R. K., and Krishnaiah, U. A. (1979), "P-stable Methods for Periodic IVPs of Second Order Differential Equations," *BIT* **19**, 347-355.

Jeltsch, R. (1976), "Note on A-stability of Multistep-Multiderivative Methods," *BIT* **16**, 74-78.

Jeltsch, R. (1978), "Complete Characterization of Multistep Methods with an Interval of Periodicity for Solving $y'' = f(x, y)$," *Mathematics of Computation* **32**, 1108-1114.

Jeltsch, R. (1979), "A-stability of Brown's Multistep Multiderivative Methods," *Numerische Mathematik* **32**, 167-181.

Jeltsch, R., and Kratz, L. (1978), "On the Stability Properties of Brown Multistep Multiderivative Methods," *Numerische Mathematik* **30**, 31-38.

Jeltsch, R., and Nevanlinna, O. (1984), "Dahlquist First Barrier for Multistage Multistep Formulas," *BIT* **24**, 538-555.

Kamel, M. S. (1981), "Improving the Efficiency of Stiff ODE Solvers by Partitioning," *Technical Report No.* **149/81**, Department of Computer Science, University of Toronto, Canada.

Kamke, E. (1943), Differential geichungen Losungsmethoden und Losungen, Leipsig: Akademische Verlagsgesellschaft.

Kaps, P., and Rentrop, P. (1979), "Generalized R-K Methods of Order Four with Stepsize Control for Stiff ODEs," *Numerische Mathematik* **33**, 55-68.

Kaps, P., and Wanner, G. (1981), "A Study of Rosenbrock-type Methods of High Order," *Numerische Mathematik* **38**, 279-298.

Kaps, P., Poon, S., and Bui, T. D. (1985), "Rosenbrock Methods for Stiff ODEs: A Comparison of Richardson Extrapolation and the Embedding Technique," *Computing* **34**, 17-40.

Kaps, P., and Osterman, A. (1985a), "Rosenbrock Methods Using Few Decompositions," *Technical Report No.* **3**, Institute fur Mathematiks und Geometrie, University of Innsbruck, FRG.

Kaps, P., and Ostermann, A. (1985b), "$L(\alpha)$-stable Variable Order Rosenbrock Methods," *Technical Report No.* **4** Institute fur Mathematiks und Geometrie, Innsbruck, FRG.

Kaps, P., and Rentrop, P. (1984), "Application of a Variable Order Semi-implicit RK Method to Chemical Models," *Computers and Chemical Engineering* **8**, 393-396.

Kaps, P. (1986), "The Solution of Combustion Problems with SIRK Methods," ODE Conference held at Sandia National Lab. Albuquerque, New Mexico, USA, July 1986.

Keller, H. B. (1968), *Numerical Methods for Two-point Boundary Value Problems,* London: Blaisdell.

Keller, H. B. (1976), "Numerical Solution of Two Point BVP," Regional Conference in Applied Mathematics, Society for Industrial and Applied Mathematics, Philadelphia.

Khaliq, A.Q.M., and Twizell, E. H. (1986), "Global Extrapolation of P-stable Methods for Second Order Periodic IVPs," ODE Conference held at Sandia National Lab., Albuquerque, New Mexico, USA, July 1986.

King, R. (1966), "Runge-Kutta Methods with Constrained Minimum Error Bounds," *Mathematics of Computation* **2**, 386-391.

Kolmogorov, A. N., and Fomin, S. V. (1957), *Elements of the Theory of Functions and Functional Analysis,* **Vol. 1**, Metric and Normed Spaces, Rochester: Graylock Press.

Kong, A. (1977), "A Search for Better Linear Multistep Methods for Stiff Problems," *Report No.* **UIUCDCS-R-77-899**, Department of Computer Science, University of Illinois at Urbana-Champaign.

Krogh, F. T. (1966), "A Predictor-corrector Method of High Order with Improved Stability Characteristics," *Journal of the ACM* **13**, 374-385.

Krogh, F. T. (1970), "VODQ/SVDQ/DVDQ--Variable Order Integrators for the Numerical Solution of ODEs," *Report No.* **CP-2308, NPO-11643**, Jet Propulsion Lab., Pasadena, California.

Krogh, F. T. (1973a), "Algorithms for Changing the Stepsize," *SIAM Journal on Numerical Analysis* **10**, 949-965.

Krogh, F. T. (1973b), "On Testing a Subroutine for Numerical Integration of ODEs," *Journal of the ACM* **20**, 545-562.

Krogh, F. T. (1975), "Changing Stepsize in the Integration of Differential Equations Using Modified Divided Differences," *Lecture Notes in Mathematics* **362**, Berlin: Springer-Verlag, 22-71.

Krogh, F. T. (1979), "Recurrence Relations for Computing with Modified Divided Differences," *Mathematics of Computation* **33**, 1265-1271.

Krogh, F. T. (1982), "Notes on Partitioning in the Solution of Stiff Equations," *Report No.* **488**, Jet Propulsion Lab., Pasadena, California.

Krogh, F. T., and Stewart, K. (1983), "Implementation of Variable Step BDF Methods for Stiff ODEs," Computer Memorandum, Jet Propulsion Lab., Pasadena, California.

Krogh, F. T., and Stewart, K. (1984), "A Symptotic ($h \rightarrow \infty$) Stability for BDFs Applied to Stiff Equations," *ACM Transactions on Mathematical Software* **10**, 45-57.

Krogh, F. T. (1986), ''What I Would Put in an ODE Solver,'' ODE Conference held at Sandia National Lab., Albuquerque, New Mexico, USA, July 1986.

Kutta, W. (1901), ''Beitrag zur Naherungs-weissen Integration totaken Differential-gleichungen,'' *Z. Math. Phys.* **46**, 435-453.

Lambert, J. D., and Shaw, B. (1965), ''On the Numerical Solution of $y' = f(x, y)$ by a Class of Formulae Based on Rational Approximations,'' *Mathematics of Computation* **19**, 456-462.

Lambert, J. D. (1971), ''Predictor-Corrector Algorithms with Identical Regions of Stability,'' *SIAM Journal on Numerical Analysis* **8**, 337-344.

Lambert, J. D. (1973a), *Computational Methods in ODEs,* New York: John Wiley.

Lambert, J. D. (1973b), ''Nonlinear Methods for Stiff Systems of ODEs,'' Proceedings of the Conference on the Numerical Solution of Differential Equations, *Lecture Notes* (G. A. Watson, ed.), Berlin: Springer-Verlag.

Lambert, J. D. (1979), ''Safe Point Methods for Separably Stiff Systems of ODEs,'' *Report No.* **NA/31**, Department of Mathematics, University of Dundee, UK.

Lambert, J. D. (1980), ''Stiffness,'' *Proceedings Computational Techniques for ODEs,* (I. Gladwell and D. K. Sayers, eds.), London: Academic Press, 19-46.

Lambert, J. D. (1987), ''Developments in Stability Theory for ODEs,'' *State of the Art in Numerical Analysis* (A. Iserles and M.J.D. Powell, eds.), Oxford: Oxford University Press, 409-431.

Lambert, J. D., and Shaw, B. (1966), ''A Method for the Numerical Solution of $y' = f(x, y)$ Based on a Self-adjusting Nonpolynomial Interpolant,'' *Mathematics of Computation* **20**, 11-20.

Lambert, J. D., and Siggurdsson, S. T. (1972), ''Multistep Methods with Variable Matrix Coefficients,'' *SIAM Journal on Numerical Analysis* **9**, 715-733.

Lambert, J. D., and Watson, I. A. (1976), ''Symmetric Multistep Methods for Periodic Initial Value Problems,'' *Journal of the Institute of Mathematics and Its Applications* **18**, 189-202.

Laurent, P. J. (1963), ''Convergence du procédé d'extrapolation de Richardson,'' *Troisième Congrès de l'AFCALTI,* Toulouse, 81-98.

Lawson, J. D. (1966), ''An Order Five R-K Process with Extended Region of Absolute Stability,'' *SIAM Journal on Numerical Analysis* **3**, 593-597.

Lawson, J. D. (1967a), ''Generalized R-K Processes for Stable Systems with Large Lipschitz Constants,'' *SIAM Journal on Numerical Analysis* **4**, 372-380.

Lawson, J. D. (1967b), ''An Order Six R-K Process with Extended Region of Stability,'' *SIAM Journal on Numerical Analysis* **4**, 620-625.

Leaf, G. K., Minkoff, M., Byrne, G. D., Sorensen, D., Bleakney, T., and Saltzman, J. (1978), "DISPL: A Software Package for One- and Two-Spatially Dimensional Kinetics-Diffusion Problems," *Report No.* **ANL-77-12**, Revision 1, Argonne National Lab.

Lee, H. B. (1967), "Matrix filtering as an aid to numerical integration," *Proceedings IEEE* **55**, 1826-1831.

Lee, D. (1974), "Nonlinear Multistep Methods for Solving IVPs in ODEs," *Report No.* **4775**, Naval Underwater Systems Center, New London Lab., Connecticut.

Lee, D. (1976), "Numerical Solution of IVPs," *Report No.* **5341**, Systems Analysis Department, Naval Underwater Systems Center, Connecticut.

Lee, D., and Preiser, S. (1977), "A Class of Nonlinear Multistep A-stable Numerical Methods for Solving Stiff DE," *Computers and Mathematics with Applications* **4**, 43-51.

Lee, H. B. (1967), "Matrix Filtering as an Aid to Numerical Integration," *Proceedings IEEE* **55** 11, 1826-1831.

Leimkuhler, B. J. (1986), "Error Estimates for DAE's," ODE Conference held at Sandia National Lab., Albuquerque, New Mexico, USA, July 1986.

Lentini, M., and Pereyra, V. (1979), "PASVA3: An Adaptive Finite Difference Program for First Order, Nonlinear Ordinary Boundary Problems," *Lecture Notes in Computer Science* **76** (B. Childs, ed.), New York: Springer-Verlag.

Lindberg, B. (1971a), "On Smoothing and Extrapolation for the Trapezoidal Rule," *BIT* **11**, 29-52.

Lindberg, B. (1971b), "On Smoothing for the Trapezoidal Rule, An Analytical Study of Some Representative Test Examples," *Report No.* **NA71.31**, Department of Science, Royal Institute of Technology, Stockholm.

Lindberg, B. (1972), "IMPEX--A Program Package for Solution of Systems of Stiff Differential Equations," *Report No.* **NA:72.50**, Department of Information Processing, Royal Institute of Technology, Stockholm.

Lindberg, B. (1973), "IMPEX 2--A Procedure for Solution of Systems of Stiff Differential Equations," *Report No.* **TRITA-NA-7303,** Department of Information Processing, Royal Institute of Technology, Stockholm.

Liniger, W. (1957), *Zur Stabilitat der numerichen Integrationsmethoden fur Differentialgleichungen,* Ph.D Thesis, University of Lausanne.

Liniger, W. (1969), "Global Accuracy and A-Stability of One and Two Step Integration Formulae for Stiff ODEs," Proceedings of the Conference on Ordinary Differential Equations, Dundee 1968, Berlin: Springer-Verlag, 188-193.

Liniger, W. (1975), "Connections Between Accuracy and Stability Properties of Linear Multistep Formulas," *Communications of the ACM* **18**, 53-56.

Liniger, W., and Willoughby, R. A. (1970), "Efficient Integration Methods for Stiff Systems of ODEs," *SIAM Journal on Numerical Analysis* **7**, 47-66.

Liniger, W., and Odeh, F. (1972), "A-stable Accurate Averaging of Multistep Methods for Stiff Differential Equations," *IBM Research Journal* **16**, Yorktown Heights.

Lotkin, M. (1951), "On the Accuracy of R-K Methods," *MTAC* **5**, 128-132.

Luke, Y. L., Fair, W., and Wimp, J. (1975), "Predictor Corrector Formulas Based on Rational Interpolants," *Journal on Computers and Mathematics with Applications* **1**, 3-12.

Mace, D., and Thomas, L. H. (1960), "An Extrapolation Method for Stepping the Calculations of the Orbit of an Artificial Satellite Several Revolutions Ahead at a Time," *Astronomical Journal* **65**, 1280.

Machura, M., and Sweet, R. A. (1980), "A Survey of Software for Partial Differential Equations," *ACM Transactions on Mathematical Software* **6**, 461-488.

Mack, I. M. (1986), "Block Implicit One-step Methods for Solving Smooth and Discontinuous DAE's," ODE Conference held at Sandia National Lab., Albuquerque, New Mexico, USA, July 1986.

MacMillan, D. B. (1968), "Asymptotic Methods for Systems of DEs in which some Variables have very short Response Times," *SIAM Journal on Applied Mathematics* **16**, 704-721.

Maeder, A. J. (1986), "A General Purpose ODE Solver Implementation," ODE Conference held at Sandia National Lab., Albuquerque, New Mexico, USA, July 1986.

Mannshardt, R. (1978), "One-Step Methods of any Order for ODEs with Discontinuous Right Hand Sides," *Numerische Mathematik* **31**, 131-152.

Matheij, R.M.M (1986), "Initial Value Methods For BVP's", ODE Conference held at Sandia National Lab., Albuquerque, New Mexico, USA, July 1986.

Merluzzi, P., and Brosilaw, C. (1978), "Runge-Kutta Integration Algorithms with Built-in Estimates of the Accumulated Truncation Error," *Computing* **20**, 1-16.

Miller, J.C.P. (1946), "The Airy Integral," in *British Association for the Advancement of Science Mathematics Table,* Cambridge: Cambridge University Press.

Miller, J.J.H. (1971), "On the Location of Zeros of Certain Classes of Polynomials with Applications to Numerical Analysis," *Journal of the Institute of Mathematics and Its Applications* **8**, 397-406.

Milne, W. E. (1953), *Numerical Solution of Differential Equations,* New York: John Wiley and Sons.

Minorsky, N. (1974), *Nonlinear Oscillations,* Huntington, NY: Robert E. Kreiger Publishing Co.

Miranker, W. L., and Wahba, G. (1976), "An Averaging Method for the Stiff Oscillatory Problem," *Mathematics of Computation* **30**, 383-399.

Miranker, W. L., Van Veldhuizen, M., and Wahba, G. (1976), "Two Methods for the Stiff Highly Oscillatory Problem," *Proceedings Conference on Numerical Analysis* (J.J.H. Miller, ed.), London: Academic Press, 257-273.

Miranker, W. L., and Van Veldhuizen, M. (1978), "The Method of Envelopes," *Mathematics of Computation* **32**, 453-496.

Moore, R. E. (1966), *Interval Analysis,* Englewood Cliffs: Prentice-Hall.

Moore, R. E. (1975), *Mathematical Elements of Scientific Computing,* New York: Holt, Rinehart and Winston.

Mortezaie, S. (1986), "Extended Adams Methods for Integration of Nonstiff IVPs," ODE Conference held at Sandia National Lab., Albuquerque, New Mexico, USA, July 1986.

Moulton, F. R. (1926), *New Methods in Exterior Ballistics,* Chicago: University of Chicago Press.

Nakhle, M., and Roux, P. (1983), "Integration of Nonlinear Systems of Stiff Differential Equations with Discontinuities," *Proceedings 10th IMACS World Congress on System Simulation and Scientific Computation,* New York, 13-16.

Neta, B. (1986), "Special Methods for Problems whose Oscillatory Solution is Damped," ODE Conference held at Sandia National Lab., Albuquerque, New Mexico, USA, July 1986.

Neta, B., and Ford, C. H. (1984), "Families of Methods for Ordinary Differential Equations Based on Trigonometric Polynomials," *Journal on Computational and Applied Mathematics* **10**, 33-38.

Nevanlinna, O., and Liniger, W. C. (1978, 1979), "Contractive Methods for Stiff Differential Equations," Parts I and II, *BIT* **18**, 457-474; *BIT* **19**, 53-72.

Neville, E. H. (1934), "Iterative Interpolation," *Journal of the Indian Mathematical Society* **20**, 87-120.

Niekerk, F.D.V. (1987), "Nonlinear One-Step Methods for IVPs," *Computers and Mathematics with Applications* **4**, 367-371.

Nordsieck, A. (1962), "On the Numerical Integration of ODEs," *Mathematics of Computation* **16**, 22-49.

Norman, A. C. (1973), *Taylor User's Manual,* Computing Lab., University of Cambridge, England.

Norman, A. C. (1976), "Expanding the Solution of Implicit Sets of Ordinary Differential Equations in Power Series," *Computer Journal* **19**, 63-68.

Norsett, S. P. (1969), "A Criterion for A(α)-stability of Linear Multistep Methods," *BIT* **9**, 259-263.

Norsett, S. P. (1974a), "Multiple Padé Approximations to the Exponential Function," *Report No.* **4/74a**, Department of Mathematics, University of Trondheim, Norway.

Norsett, S. P. (1974b), "Semi-explicit R-K Methods," *Report No.* **6/74b**, Department of Mathematics, University of Trondheim, Norway.

Norsett, S. P. (1974c), "One Step Methods of Hermite Type for Numerical Integration of Stiff Systems," *BIT* **14**, 63-77.

Norsett, S. P. (1975), "C-polynomials for Rational Approximation to the Exponential Function," *Numerische Mathematik* **25**, 39-56.

Norsett, S. P. (1978), "Restricted PadéApproximation to the Exponential," *SIAM Journal on Numerical Analysis* **15**, 1008-1029.

Norsett, S. P. (1986), "Parallel R-K Methods," ODE Conference held at Sandia National Lab., Albuquerque, New Mexico, USA, July 1986.

Norsett, S. P., and Wolfbrandt, A. (1979), "Order Conditions for Rosenbrock Type Methods," *Numerische Mathematik* **32**, 1-15.

Norsett, S. P., and Thomsen, P. G. (1984), "Embedded SDIRK Methods of Basic Order Three," *BIT* **24**, 634-646.

Norsett, S. P., and Thomsen, P. G. (1986a), "Local Error Control in SDIRK," *BIT* **26**, 100-113.

Norsett, S. P., and Thomsen, P. G. (1986b), "SIMPLE, a Stiff System Solver," ODE Conference held at Sandia National Lab., Albuquerque, New Mexico, USA, July 1986.

Nowak, U., and Deuflhard, P. (1985), "Numerical Identification of Selected Rate Constants in Large Chemical Reaction Systems," *Journal on Applied Numerical Mathematics* **1**, 59-75.

Nystrom, E. J. (1925), "Ueber Die Numerische Integration von Differential-gleichungen," *Acta Soc. Sc. Fenn.* **50** 13, 1-55.

Odeh, L. (1971), "An Experimental and Theoretical Analysis of SAPS Method for Stiff ODEs," *Report No.* **NA71.28**, Department of Information Processing, Royal Institute of Technology, Stockholm.

Odeh, F., and Liniger, W. (1975), "Nonlinear Fixed-h Stability of Multistep Methods," *Report No.* **RC5717**, IBM, Yorktown Heights.

O'Regan, P. G. (1970), "Stepsize Adjustment at Discontinuities for Fourth Order R-K Methods," *Computer Journal* **13**, 401-404.

Ortega, J. M., and Rheinboldt, W. (1970), *Iterative Solution of Nonlinear Equations in Several Variables,* New York: Academic Press.

Ortega, J. M., and Voigt, R. G. (1985), "Solution of Partial Differential Equations on Vector and Parallel Computers," *SIAM Review* **27**, 149-240.

Ostermann, A., Kaps, P., and Bui, T. D. (1986), "The Solution of a Combustion Problem with Rosenbrock Methods," ODE Conference held at Sandia National Lab., Albuquerque, New Mexico, USA, July 1986.

Petzold, L. (1981), "An Efficient Numerical Method for Highly Oscillatory ODEs," *SIAM Journal on Numerical Analysis* **18**, 455-479.

Petzold, L. (1982a), "A Description of DASSL: A Differential/Algebraic System Solver," *IMACS Transactions on Scientific Computation*, 430-432.

Petzold, L. (1982b), "Differential/Algebraic Equations are not ODEs," *SIAM Journal on Scientific and Statistical Computing* **3**, 367-384.

Petzold, L. (1983), "Automatic Selection of Method for Solving Stiff and Nonstiff Systems of ODEs," *Report No.* **SAND80-8230**, Sandia National Lab., Livermore, California; *SIAM Journal on Scientific and Statistical Computing* **4**, 136-148.

Petzold, L. (1984), "Order Results for IRK Methods Applied to differential Algebaic Systems," *Report No.* \fBSAND84-8838\fR, Sandia National Lab., Albuquerque, New Mexico, USA.

Petzold, L. (1986), "Numerical Methods For Differential Algebraic Systems," ODE Conference held at Sandia National Lab., Albuquerque, New Mexico, USA, July 1986.

Philips, E. G. (1962), *A Course of Analysis*, Cambridge: Cambridge University Press.

Piotrowski, P. (1969), "Stability, Consistency and Convergence of Variable K-step Methods for Numerical Integration of Large Systems of ODEs," *Proceedings Conference on Numerical Solution of ODEs, Dundee* (J. L. Morris, ed.), New York: Springer-Verlag, 221-227.

Pontryagin, L. S. (1962), *Ordinary Differential Equations*, Reading, Massachusetts: Addison-Wesley Publishing Co., Inc.

Prince, P. J., and Dormand, J. R. (1981), "High Order Embedded RKF's," *Journal on Computational and Applied Mathematics* **7**, 67-75.

Prothero, A., and Robinson, A. (1974), "On the Stability and Accuracy of One Step Methods for Solving Systems of ODEs," *Mathematics of Computation* **28**, 145-162.

Prothero, A. (1980), "Estimating the Accuracy of Numerical Solution to ODEs," *Proceedings Computational Techniques for ODES* (I. Gladwell and D. K. Sayers, eds.), London: Academic Press, 103-128.

Prothero, A., and Robinson, A (1974), "On the Stability and Accuracy of One-step Methods for Solving Stiff Systems of Ordinary Differential Equations," *Mathematics of Computation* **28**, 145-162.

Rall, L. B. (1981), *Automatic Differentiation: Techniques and Applications, Lecture Notes in Computer Science*, Vol. **120**, New York: Springer-Verlag.

Ralston, A. (1962), "R-K with Minimum Error Bounds," *Mathematics of Computation* **16**, 431-437.

Ralston, A. (1965a), "Relative Stability in the Numerical Solution of ODEs," *SIAM Review* **7**, 114-125.

Ralston, A. (1965b), *A First Course in Numerical Analysis,* New York: McGraw-Hill.

Richardson, L. F. (1927), "The Deferred Approach to the Limit, I--Single Lattice," *Transactions of the Royal Society of London* **226**, 299-349.

Richtmyer, R., and Morton, K. W. (1967), *Difference Methods for IVPs,* New York: Interscience Publishers, second edition.

Robertson, H. H. (1976), "Numerical Integration of Systems of Stiff ODEs with Special Structure," *Journal of the Institute of Mathematics and Its Applications* **18**, 249-263.

Robinson, A., and Prothero, A. (1977), "Global Error Estimates for Solution of Stiff Systems of ODEs," *Dundee Conference on Numerical Analysis,* New York: Springer-Verlag.

Rolfes, L. (1981), "A Global Method for Solving Stiff Differential Equations," *NRIMS Special Report* **TWISK 228 CSIR**, Pretoria.

Rolfes, L., and Snyman, J. A. (1982), "An Evaluation of a Global Method Applied for Stiff ODEs," Report, University of Pretoria.

Romberg, W. (1955), "Vereinfachte Numerische Integration," *Norske Vid. Selsk. Forhandlinger(* Trondheim) **28**, 30-36.

Rosenbrock, H. H. (1963), "Some General Implicit Processes for the Numerical Solution of Differential Equations," *Computer Journal* **5**, 329-330.

Rosser, J. B. (1967), "A Runge-Kutta for All Seasons," *SIAM Review* **9**, 417-452.

Runge, C. (1895), "Ueber Die Numerische Aufoesung von Differentialgeichungen," *Math. Ann.* **46**, 167-178.

Rutishauser, H. (1960), "Bemerkungen zur Numerischen Integration Gewohlicher Differential-gleichungen n-ter Ordnung Numerische," *Numerische Mathematik* **2**, 263-279.

Sacks-Davis, R. (1977), "Solution of Stiff ODEs by a Second Derivative Method," *SIAM Journal on Numerical Analysis* **14**, 1088-1100.

Sacks-Davis, R. (1980), "Fixed Leading Coefficient Implementation of Second Derivative Formulas for Stiff ODEs," *ACM Transactions on Mathematical Software* **6**, 540-562.

Sacks-Davis, R., and Shampine, L. F. (1981), "A Type Insensitive ODE Code Based on Second Derivative Formulas," *Computations and Mathematics with Applications* **7**, 487-495.

Salane, D. E. (1986), "A Code for Solving Large Sparse Systems of Stiff Ordinary Differential Equations," ODE Conference held at Sandia National Lab., Albuquerque, New Mexico, USA, July 1986.

Sanchez, D. (1968), *Ordinary Differential Equations and Stability Theory: An Introduction,* New York: W. H. Freeman and Co.

Sand, J. (1979), "Description of The SOM System," *Report No.* **ISSSN 0105-8525**, Department of Computer Science, Aarhus University, Denmark.

Sand, J., and Østerby, O. (1979), "Regions of Absolute Stability," *Report No.* **0105-8517**, Department of Computer Science, Aarhus University, Denmark.

Sand, J. (1981), "On One-leg and Linear Multistep Formulas with Variable Stepsizes," *Report No.* **8112**, Royal Institute of Technology, Stockholm.

Sarafyan, D. (1965), "Multistep Methods for the Numerical Solution of ODEs Made Self-Starting," Technical Report No. **495**, Mathematics Research Center, Madison, Wisconsin.

Scheid, R. E. (1984), "The Accurate Numerical Solution of Highly Oscillatory ODEs," *Mathematics of Computation* **41**, 487-509.

Scraton, R. E. (1964), "Estimation of the Truncation Error--R-K and Allied Processes," *Computer Journal* **7**, 246-248.

Sedgwick, A. E. (1973), "An Effective Variable Order, Variable Stepsize Adams Method," Ph.D. Dissertation, *Technical Report* **53**, Department of Computer Science, University of Toronto, Canada.

Shampine, L. F. (1972a), "Stability Regions for Extrapolated R-K and Adams Methods," *Technical Report* **SC-RR-72-0223**, Sandia National Lab., Albuquerque, New Mexico.

Shampine, L. F. (1972b), "Some Numerical Experiments with DIFSUB," *ACM SIGNUM Newsletter* **7**, 24-26.

Shampine, L. F. (1973), "Local Extrapolation in the Solution of ODEs," *Mathematics of Computation* **27**, 91-97.

Shampine, L. F. (1975), "Stiffness and Non-stiff Differential Equation Solvers," in *Numerische Behandlung von Differential-gleichungen* (L. Collatz, ed.), *International Series Numerical Mathematics* **27**, Basel, Switzerland: Birkhauser, 287-301.

Shampine, L. F. (1977a), "Starting an ODE Solver," *Report No.* **77-1023**, Sandia National Lab., Albuquerque, New Mexico.

Shampine, L. F. (1977b), "Local Error Control in Codes for ODEs," *Applied Mathematical Computation* **3**, 189-210.

Shampine, L. F. (1979), "Evaluation of Implicit Formulas for the Solution of ODEs," *BIT* **19**, 495-502.

Shampine, L. F. (1980), "Lipschitz Constants and Robust ODE Codes," Conference on Computational Methods in Nonlinear Mechanics (J. T. Oden, ed.), Amsterdam: North-Holland Publishing Co., 427-449.

Shampine, L. F. (1981a), "Evaluation of a Test Set for Stiff ODE Solvers," *ACM Transactions on Mathematical Software* **7**, 409-420.

Shampine, L. F. (1981b), "Type Insensitive ODE Codes Based on Implicit A-stable Formulas," *Mathematics of Computation* **36** 154, 499-510.

Shampine, L. F. (1982), "Implementation of Rosenbrock Methods," *ACM Transactions on Mathematical Software* **8**, 93-113.

Shampine, L. F. (1983a), "Global Error Estimation for Stiff ODEs," *Report No.* **SAND82-2517**, Sandia National Lab., Albuquerque, New Mexico;

Shampine, L. F. (1983b), "Efficient Extrapolation Methods for ODEs I," *IMA Journal on Numerical Analysis* **3**, 383-395.

Shampine, L. F. (1983c), "Type Insensitive ODE Codes Based on Extrapolation Methods," *SIAM Journal on Scientific and Statistical Computing* **4**, 635-664.

Shampine, L. F. (1983d), "Efficient Extrapolation Methods for ODEs II," *Report No.* **SAND-83-1927**, Sandia National Lab., Albuquerque, New Mexico.

Shampine, L. F. (1984a), "Asymptotic Bounds on the Errors of One-step Methods," *Numerische Mathematik* **45**, 201-206.

Shampine, L. F. (1984b), Private Communication.

Shampine, L. F. (1984c), "Fixed vs. Variable Order Runge-Kutta," *Report No.* **SAND-84-1410**, Sandia National Lab., Albuquerque, New Mexico.

Shampine, L. F. (1984d), "Local Error Estimation by Doubling," *Report No.* **SAND-84-0413**, Sandia National Lab., Albuquerque, New Mexico.

Shampine, L. F. (1985a), "Control of Step Size and Order in Extrapolation Codes," *Report No.* **SAND-85-0725**, Sandia National Lab., Albuquerque, New Mexico.

Shampine, L. F. (1985b), "The Stepsize used by One-Step Codes for ODEs," *Numerische Mathematik* **1**, 95-106.

Shampine, L. F. (1985c), "Interpolation for Runge-Kutta Methods," *SIAM Journal on Numerical Analysis* **22**, 1014-1027.

Shampine, L. F. (1986), "Some Practical RKF's," *Mathematics of Computation* **46**, 135-150.

Shampine, L. F., and Allen, R. C. (1973), *Numerical Computing: An Introduction,* Philadelphia: W. B. Sanders Publishers.

Shampine, L. F., and Baca, L. S. (1983), "Smoothing the Extrapolated Midpoint Rule," *Numerische Mathematik* **41**, 165-175.

Shampine, L. F., and L. S. Baca (1985), "Global Error Estimates for ODEs Based on Extrapolation Methods," *SIAM Journal on Scientific and Statistical Computing* **6**, 1-14.

Shampine, L. F., Baca, L. S., and Bauer, H. J. (1983), "Output in Extrapolation Codes," *Computation and Mathematics with Applications* **9**, 245-255.

Shampine, L. F., and Gordon, M. K. (1975), *Computer Solution of ODEs,* San Francisco: W. H. Freeman and Co.

Shampine, L. F., and Watts, H. A. (1971), "Comparing Error Estimators for R-K Methods," *Mathematics of Computation* **25**, 445-455.

Shampine, L. F., and Watts, H. A. (1976a), "Global Error Estimation for ODEs," *ACM Transactions on Mathematical Software* **2**, 172-186.

Shampine, L. F., and Watts, H. A. (1976b), "CACM Algorithm 504, GERK: Global Estimation for ODEs," *ACM Transactions on Mathematical Software* **2**, 200-203.

Shampine, L. F., and Watts, H. A. (1976c), "Practical Solution of ODEs by Runge-Kutta Methods," *Report No.* **SAND-76-0585**, Sandia National Lab., Albuquerque, New Mexico.

Shampine, L. F., and Watts, H. A. (1977), "The Art of Writing a Runge-Kutta Code, Part I," *Math. Software III* (J. R. Rice, ed.), New York: Academic Press.

Shampine, L. F., and Watts, H. A. (1979a), "DEPAC--Design of a User Oriented Package of ODE Solvers," *Report No.* **SAND79-2374**, Sandia National Lab., Albuquerque, New Mexico.

Shampine, L. F., and Watts, H. A. (1979b), "The Art of Writing R-K Code II," *Applied Mathematics of Computation* **5**, 93-121.

Shampine, L. F., Watts, H. A., and Davenport, S. M. (1976), "Solving Non-stiff ODEs—The State of the Art," *SIAM Review* **18**, 376-411.

Shaw, B. (1967), "Some Multistep Formulae for Special High Order ODEs," *Numerische Mathematik* **9**, 367-378.

Shintani, H. (1966a), "On a One-step Method of Order Four," *J. Science Hiroshima University Ser. A-I Math.* **30**, 91-107.

Shintani, H. (1966b), "Two Step Processes for One-step Methods of Order 3 and of Order 4," *J. Science Hiroshima University Ser. A-I Math.* **30**, 183-195.

Shinad, H. (1980), "Global Error Estimation for the Implicit Trapezoidal Rule," *BIT* **20**, 120-121.

Skeel, R. D. (1973), "Convergence of Multivalue Methods for Solving ODEs," *Report No.* **TR73-16**, Department of Computing Sciences, University of Alberta, Canada.

Skeel, R. D. (1976), "Analysis of Fixed Stepsize Methods," *SIAM Journal on Numerical Analysis* **13**, 664-685.

Skeel, R. D., and Jackson, L. W. (1977), "Consistency of Nordsieck Methods," *SIAM Journal on Numerical Analysis* **14**, 910-924.

Skeel, R. D., and Kong, A. K. (1977), "Blended Linear Multistep Methods," *ACM Transactions on Mathematical Software* **3**, 326-345.

Skeel, R. D. (1979a), "Equivalent Forms of Multistep Formulas," *Mathematics of Computation* **33**, 1229-1250.

Skeel, R. D. (1979b), "An Incomplete List of Better Fortran Codes for IVPs in ODEs," *ACM SIGNUM Newsletter* **14**, 22-24.

Skeel, R. D. (1982), "A Theoretical Framework for Proving Accuracy Results for Deferred Corrections," *SIAM Journal on Numerical Analysis* **19**, 171-196.

Skeel, R. D., and Jackson, L. W. (1983), "The Stability of Variable-Stepsize Nordsieck Methods," *SIAM Journal on Numerical Analysis* **20**, 840-853.

Skeel, R. D. (1984), "Construction of Variable Stepsize Multistep Formulas," manuscript, Department of Computer Science, University Illinois at Urbana-Champaign.

Skeel, R. D. (1985), "Computational Error Estimates for Stiff ODEs," *Proceedings International Conference on Computational Mathematics*, University of Benin, Nigeria, November 2-4, 1983 (S. O. Fatunla, ed.), Dublin, Ireland: Boole Press, 3-10.

Skeel, R. D. (1986a), "Thirteen Ways to Estimate Global Errors," *Numerische Mathematik* **48**, 1-20.

Skeel, R. D. (1986b), "Global Error Estimation and the BDF's," ODE Conference held at Sandia National Lab., Albuquerque, New Mexico, USA, July 1986.

Skeel, R. D. (1986c), "Global Error Estimation and the Backward Differentiation Formulas," *Report No.* **UIUCDCS-R-86-1291**, Department of Computer Science, University Illinois at Urbana-Champaign.

Skelboe, S. (1986), "Stability Properties of Linear Multirate Formulas," ODE Conference held at Sandia National Lab., Albuquerque, New Mexico, USA, July 1986.

Sloate, H., and Bickart, T. (1973), "A-stable Composite Multistep Methods," *Journal of the ACM* **20**, 7-26.

Sommeijer, B. P. (1986), "On Economization of Explicit RK Methods," *Journal on Applied Numerical Mathematics* **2**, 57-86.

Sommeijer, B. P., Van der Houwen, P. J., and Neta, B. (1986), "Symmetric Linear Multistep Methods for Second Order Differential Equations with Periodic Solutions," *Journal on Applied Numerical Mathematics* **2**, 69-77.

Starner, J. W. (1976), "A Numerical Algorithm for the Solution of Implicit Algebraic-differential Systems of Equations," *Technical Report No.* **318**, Department of Mathematics and Statistics, University of New Mexico, Albuquerque, New Mexico.

Steihaug, T., and Wolfbrandt, A. (1979), "An Attempt to Avoid Exact Jacobian and Nonlinear Equations in the Numerical Solution of Stiff Differential Equations," *Mathematics of Computation* **33**, 521-534.

Stetter, H. J. (1965a), "A Study of Strong and Weak Stability in Discretization Algorithms," *SIAM Journal on Numerical Analysis* **2**, 265-280.

Stetter, H. J. (1965b), "Asymptotic Expansions for the Error of Discretization Algorithms for Nonlinear Functional Equations," *Numerische Mathematik* **7**, 18-31.

Stetter, H. J. (1969), "Stability Properties of Extrapolation Methods," *Conference on Numerical Solution of Differential Equations,,* University of Dundee (J.L. Morris, ed.), Berlin: Springer-Verlag.

Stetter, H. J. (1971), "Local Estimation of the Global Discretization Error," *SIAM Journal on Numerical Analysis* **8**, 512-52.

Stetter, H. J. (1973), *Analysis of Discretization Methods for ODEs,* Berlin: Springer-Verlag.

Stetter, H. J. (1978), "The Defect Correction Principle and Discretization Methods," *Numerische Mathematik* **29**, 425-443.

Stetter, H. J. (1979a), "Global Error Estimation in ODE Solvers," *Lecture Notes in Mathematics* **630**, Berlin: Springer-Verlag, 179-189.

Stetter, H. J. (1979b), "Global Error Estimation in Adams PC-Codes," *ACM Transactions on Mathematical Software* **5**, 415-430.

Stetter, H. J. (1979c), "Interpolation and Error Estimation in Adams PC-codes," *SIAM Journal on Numerical Analysis* **16**, 311-323.

Stetter, H. J. (1980), "Tolerance Proportionality in ODE-codes," *Seminarberichte Nr. 32 Sektion Math.* (R. Marz ed.), Berlin: Humboldt University, 109-123.

Stetter, H. J. (1981), "On the Error Control in ODE Solvers with Local Extrapolation," *Computing* **27**, 169-177.

Stetter, H. J. (1982), "Global Error Estimation in Ordinary IVPs," *Lecture Notes in Mathematics* **968**, Berlin: Springer-Verlag.

Stetter, H. J. (1986) "Enclosing Solutions of Ordinary IVPs Computationally," ODE Conference held at Sandia National Lab., Albuquerque, New Mexico, USA, July 1986.

Stewart, K., and Krogh, F. T. (1986) "Semi-implicit BDF's: Accuracy and Stability," ODE Conference held at Sandia National Lab., Albuquerque, New Mexico, USA, July 1986.

Stiefel, E., and Bettis, D. G. (1969), "Stabilization of Cowell's Method," *Numerische Mathematik* **13**, 154-175.

Stoer, J. (1961), "Ueber Zwei Algorithmen zur Interpolation mit Rationalen Funktionen," *Numerische Mathematik* **3**, 285-304.

Stoer, J. (1974), "Extrapolation Methods for the Solution of IVPs and their Practical Realisation," *Lecture Notes in Mathematics* **362**, Berlin: Springer-Verlag, 1-21.

Stummel, F., and Hainer, K. (1980), *Introduction to Numerical Analysis* (E. R. Dawson, trans.), Edinburgh: Scottish Academic Press.

Tendler, J. M., Bickart, T. A., and Picel, Z. (1978a), "A Stiffly Stable Integration Process Using Cyclic Composite Method," *ACM Transactions on Mathematical Software* **4**, 339-368.

Tendler, J. M., Bickart, T. A., and Picel, Z. (1978b), "Algorithm 534 STINT: Stiff (differential equations) Integrator," *ACM Transactions on Mathematical Software* **4**, 399-403.

Thomas, R. M. (1984), "Phase Properties of High Order, almost P-stable Formulae," *BIT* **24**, 225-238.

Thomsen, P. G. (1979), "Numerical Simulation in Theory and Practice," *Technical Report* **2**, Department of Mathematics, University of Trondheim, Norway.

Tischer, P. E., and Gupta, G. K. (1983a), "Some New Cyclic LMF for Stiff Systems," *Technical Report* **40**, Department of Computer Science, Monash University, Australia.

Tischer, P. E., and Gupta, G. K. (1983b), "A Cyclic Method for Stiff ODE Solver," *Technical Report No.* **38**, Department of Computer Science, Monash University, Australia.

Tischer, P. E., and Sacks-Davis, R. (1983), "A New Class of Cyclic Multistep Formulae for Stiff Systems," *SIAM Journal on Scientific and Statistical Computing* **4**, 733-746.

Tischer, P. E. (1984), "Propagated Error Behaviour of Multistep Formulae," *Technical Report No.* **41**, Department of Computer Science, Monash University, Australia.

Tischer, P. E., and Gupta, G. K. (1985), "An Evaluation of Some New Cyclic Multistep Formulas for Stiff Ordinary Differential Equations" *ACM Transactions on Mathematical Software* **11**, 263-270.

Twizell, E. H., and Khaliq, A.Q.M. (1981), "One-step Multiderivative Methods for First Order Ordinary Differential Equations," *BIT* **21**, 518-527.

Twizell, E. H., and Khaliq, A.Q.M. (1984), "Multiderivative Methods for Periodic IVPs," *SIAM Journal on Numerical Analysis* **21**, 111-122.

Uspensky, J. V. (1948), *Theory of Equations,* New York: McGraw-Hill.

Van der Houwen, P. J. (1979), "Stabilized R-K Methods for Second Order DEs without Derivatives," *SIAM Journal on Numerical Analysis* **16**, 523-537.

Van der Houwen, P. J., and Sommeijer, B. P. (1979), "On the Internal Stability of Explicit, M-stage R-K Methods for Large Eigenvalues," *Report No.* **NW 72/79**, Department of Numerical Mathematics, Math Centrum, Amsterdam.

Van der Houwen, P. J., and Sommeijer, B. P. (1984), "Multistep Methods with Reduced Truncation Error for Periodic IVPs," *IMA Journal on Numerical Analysis* **4**, 479-489.

Varah, J. M. (1979), "On the Efficient Implementation of Implicit R-K Methods," *Mathematics of Computation* **33**, 557-561.

Van Veldhuizen, M. (1981), "D-Stability," *SIAM Journal on Numerical Analysis* **18**, 45-64.

Verner, J. H. (1978), "Explicit R-K Methods with Estimates of the Local Truncation Errors," *SIAM Journal on Numerical Analysis* **15:4**, 772-790.

Verner, J. H. (1979), "Families of Imbedded R-K Methods," *SIAM Journal on Numerical Analysis* **16**, 857-875.

Verwer, J. G. (1981), "On the Practical Value of the Notion of BN-stability," *BIT* **21**, 355-361.

Verwer, J. G. (1981), "An Analysis of Rosenbrock Methods for Nonlinear Stiff IVPs," *SIAM Journal on Numerical Analysis* **19**, 155-170.

Verwer, J. G. (1982), "Instructive Experiments with Some R-K Rosenbrock Methods," *Computation and Mathematics with Applications* **8**, 217-229.

Voss, D. A. (1986), "Factored 2-step R-K Methods," ODE Conference held at Sandia National Lab., Albuquerque, New Mexico, USA, July 1986.

Wait, R. (1979), *The Numerical Solution of Algebraic Equations,* Chichester: John Wiley and Sons.

Wanner, G. (1977), "On the Integration of Stiff DEs," *ISNM* **37** (J. Descloux and J. Marti, eds.), Basel: Birkhauser.

Wanner, G. (1980), "On the Choice of γ for Singly Implicit R-K or Rosenbrock Methods," *BIT* **20**, 102-106.

Wanner, G. (1987), "Order Stars and Stability," *State of the Art in Numerical Analysis* (A. Iserles and M.J.D. Powell, eds.), Oxford: Oxford University Press.

Wanner, G., Hairer, E., and Norsett, S. P. (1978), "Order Stars and Stability Theorems," *BIT* **18**, 475-489.

Watanabe, D. S., and Sheikh, Q. M. (1984), "One-leg Formulas for Stiff ODEs," *SIAM Journal on Scientific and Statistical Computing* **5**, 489-496.

Watkins, D. S. (1981a), "Determining Initial Values for Stiff Systems of ODEs," *SIAM Journal on Numerical Analysis* **18**, 13-20.

Watkins, D. S. (1981b), "Efficient Initialization of Stiff Systems with One Unknown Initial Condition," *SIAM Journal on Numerical Analysis* **18**, 794-800.

Watkins, D. S., and Hanson, R. H. (1983), "The Numerical Solution of Separably Stiff Systems by Precise Partitioning," *ACM Transactions on Mathematical Software* **9**, 293-301.

Watts, H. A. (1983), "Starting Stepsize for an ODE Solver," Journal on Computational and Applied Mathematics **9**, 177-191.

Widlund, O. (1967), "A Note on Unconditionally Stable LMM," *BIT* **7**, 65-70.

Wolfbrandt, A. (1977a), "A Note on Recent Results of Rational Approximation to the Exponential Function," *BIT* **17**, 367-368.

Wolfbrandt, A. (1977b), "A Study of Rosenbrock Processes with Respect to Order Conditions and Stiff Stability," Department of Computer Science, Chalmers University of Technology and University of Goteborg, Sweden.

Wuytack, L. (1971), "A New Technique for Rational Extrapolation to the Limit," *Numerische Mathematik* **17**, 215-221.

Wuytack, L. (1974), "Extrapolation to the Limit by Using Continued Fraction Interpolation," *Rocky Mountain Journal of Mathematics* **4**, 395-397.

Zadunaisky, P. E. (1976), "On the Estimation of Errors in the Numerical Integration of ODEs," *Numerische Mathematik* **27**, 21-39.

Zennaro, M. (1986), "Stability Properties of Interpolants for R-K Methods," ODE Conference held at Sandia National Lab., Albuquerque, New Mexico, USA, July 1986.

Zhou, B. (1986) "A-Stable and L-Stable Block Implicit One-step Methods," ODE Conference held at Sandia National Lab., Albuquerque, New Mexico, USA, July 1986.

Zlatev, Z. (1978), "Stability Properties of Variable Stepsize Variable Formula Methods," *Numerische Mathematik* **31**, 175-182.

Zlatev, Z. (1981), "Modified Diagonally Implicit R-K Methods," *SIAM Journal on Scientific and Statistical Computing* **2**, 321-334.

Zlatev, Z., and Thomsen, P. G., "Automatic Solution of Differential Equations Based on the Use of LMM," *ACM Transactions on Mathematical Software* **5**, 401-414.

Zlatev, Z., Berkowicz, R., and Prahm, L. P. (1983), "Testing Subroutines Solving Advection-Diffusion Equations in Atmospheric Environments," *Computer and Fluids* **11**, 13-38.

Zlatev, Z., Wasniewski, J., and Schaumburg, K. (1985), "Exploiting the Sparsity in the Solution of Linear Ordinary Differential Equations," *Computation and Mathematics with Applications* **11**, 1069-1087.

INDEX

A

Absolute stability, 172–180
  of Euler scheme, 157
  of LMMs, 118
  with nonstiff algorithms, 167
  of one-step methods, 34–35
  with R-K methods, 51
Adams codes, 219, 241
  automatic implementation of, 119–124
  for LMM starting values, 90
  for nonstiff algorithms, 202
  for stiff zones, 167
  See also specific codes
Adams-Bashford scheme, 121
  for LMM, 94–95
  RAS for, 176
Adams-Bashford-Moulton scheme, 102,
    105–108
  RAS for, 178
  with second order differential equations,
    228
Adams-Moulton formula, 121–123
  extrapolation schemes with, 158
  for implicit LMM, 95–96
  stability polynomial for, 193
  for stiff problems, 169, 202
  for stiff/nonstiff problems, 213–214
Algebraic stability for R-K methods, 188,
    192–195
AMF. *See* Adams-Moulton formula
Angle of stability
  with BDF, 184, 212
  with MEBDF, 212
  with stiff methods, 185
AN-stability for R-K methods, 188–191
A-stability
  and AN-stability, 190–191
  and contractivity, 181
  of implicit midpoint rule, 146

  of implicit R-K schemes, 77, 188–189
  of inverse Euler scheme, 158
  of numerical integrators, 181–184
  of one-step methods, 34
  of R-K schemes, 73–74
  with semi-implicit R-K scheme, 79
  of SIRK, 208
  with stiff problems, 164, 223
  of trapezoidal schemes, 45, 144
$A(\propto)$-stability, 182–183
$A(\propto, D)$-stability, 184, 188
$A(O)$-stability, 183, 187
$A(\infty)$-stability, 184, 186

B

Backward difference operator, 1
Backward differentiation formulas, 241
  angle of absolute stability of, 185, 212
  coefficients for, 100
  costs of, compared to IRK, 204
  for DAE, 240
  for LMM starting values, 90
  for R-K methods, 195
  stability polynomial for, 193
  for stiff problems, 163, 169, 183–184,
    198, 200, 202, 211–215
Backward Euler implicit Runge-Kutta
  scheme, 76
  for stiff IVPs, 169
Backward interpolation formula, 8–9
BDF. *See* Backward differentiation
  formulas
BLEND code, 210
  angle of absolute stability of, 185
BLENDED code, for stiff problems, 208
Blended-DIFSUB code and IVP stiffness,
  27

BN-stability for R-K methods, 188–192
Boundary locus method, 179
Boundary value problems, 245
Brown methods, 179–180
B-stability
  of one-step methods, 182
  for R-K methods, 188–189
Butcher Runge-Kutta schemes
  implicit, 77
  semi-implicit, 79
BVPSOL code, 245

C

Central difference operator, 2
Classical Runge-Kutta schemes
  four-stage, 65, 74
  three-stage, matrix for, 64
Coefficients
  with explicit LMMs, 91–94
  with implicit LMMs, 95–98
  linear systems with constant, 13–15
  for multiderivative methods, 234–235
  of R-K methods, 50–51, 58
Collocation methods, 245
COLSYS code, 245
Common norms, 18–19
Composite multistep methods, 198
Consistency
  of LMMs, 113–114
  of one-step methods, 31, 33, 42
Constant coefficients, linear systems with, 13–15
Contraction mapping theorem and convergence, 169
Contractivity, 181–182, 196
Convergence
  of Adams-Moulton scheme, 123
  of discretization methods, 30
  of Euler scheme, 36–37, 39–42
  of extrapolation processes, 154–158
  of implicit R-K schemes, 75
  of iteration schemes, 168–171
  of LMMs, 104, 113–115, 188
  of one-step schemes, 33
  of R-K schemes, 62
Coupling
  and partitioning, 241
  in stiff algorithms, 200
Cyclic methods, 243
  for stiff algorithms, 164, 198

D

D02BDF code, 30
D02PAF code, 54–55
DAE. *See* Differential algebraic equations
Dahlquist Barrier Theorem, 238
  and A-stability, 164, 181
  and LMM order, 115
  and root distributions, 15
  and trapezoidal scheme stability, 45
DEBDF code, 213–215
Defect correction approach, 244–245
DEGRK code
  and R-K local error estimates, 54
  and Rosenbrock local extrapolation, 88
  for stiff algorithms, 208
DEPAC code for stiff problems, 27, 120, 202
DETEST code for stiff algorithms, 202
Diagonally-implicit Runge-Kutta scheme, 239
  coefficients for, 224
  for stiff problems, 200, 208–210
DIFEX1 code
  for R-K schemes, 55
  with singularities, 136–137
DIFEX2 code with singularities, 136, 138
Difference calculus, theorem for, 2–3
Difference equations, finite, 10–13
Difference operators, 1–3
Differential algebraic equations
  BFD for, 240
  with one-leg multistep methods, 196
DIFSUB code, 119, 238–239, 242
  partitioning with, 241
  stepsize selection problem in, 26
  for stiff problems, 27, 167–168, 202, 208, 210
  for stiff/nonstiff differential systems, 213–215
DIRK. *See* Diagonally-implicit Runge-Kutta scheme
DIRK1 code, angle of absolute stability of, 185
DIRK2 code, angle of absolute stability of, 185
Discontinuities, 125–127
  local error estimates for, 136, 139
  and stepsize selection, 26–27
  switching functions for, 242
Discrete variable methods, 22, 25
Discretization errors, 30

Distribution of roots of polynomials, 15–17
Divided difference scheme, 5–9
Doubling error estimator with R-K methods, 55
DSTIFF code
    angle of absolute stability of, 185
    for stiff problems, 211
    for stiff/nonstiff differential systems, 213
DVDQ code, 119
DVERK code, 54–55

E

Embedded methods
    L-stable, for stiff problems, 225
    for R-K local error estimates, 54, 203
Enright scheme
    boundary locus for, 179
    for discontinuities, 27
    with scalar iteration matrices, 212, 215
    with second derivative formulas, 216
EPISODE code, 119
    for stiff/nonstiff differential systems, 213
Error constants
    for LMMs, 109, 112
    for multiderivative methods, 234–235
    for Simpson's rule, 111–112
Errors and error functions, 28–30
    in discontinuous systems, 137–139
    with explicit LMMs, 91–92
    for explicit R-K two-stage process, 57, 59
    with implicit LMMs, 99
    for Lagrangian interpolation, 4
    in nonlinear multistep methods, 136
    for nonstiff algorithms, 167–168
    with one-leg multistep methods, 197
    with polynomial extrapolation, 149
    with predictors, 103
    for R-K processes, 52, 72–74
    with Rosenbrock method, 86
    with semi-implicit R-K schemes, 81
    with singularities, 130
    for SIRK, 208
    with stiff IVPs, 163–164
    See also Global errors; Local errors; Local truncation errors
Euclidean norm, 18–19

Euler schemes, 35–37
    convergence of, 39–42
    extrapolation with, 37–39
    for extrapolation tables, 142–144
    matrix representation of, 63
    with R-K two-stage process, 58
    stability of, 156–158
    See also Explicit Euler schemes; Implicit Euler schemes
Existence of solutions, 23–25
Expansion methods, 22
Explicit Euler schemes
    with discontinuities, 139
    for scalar IVPs, 162
    with stiff systems, 163
Explicit linear multistep methods, 90–95
Explicit midpoint rule
    and polynomial extrapolation, 149
    with R-K two-stage process, 58
Explicit Runge-Kutta schemes
    absolute stability with, 172–173
    coefficients for, 50, 224
    matrix for, 56–60
    for nonstiff problems, 203
    order with, 51
    stability polynomial for, 193
    for stiff problems, 166, 205
    trapezoid, 64
Exponential fitting for stiff IVPs, 164
EXPTR-1 code, 43–44
EXPTR-2 code, 43–44
EXPTR-3 code, 43–44
Extrapolation methods, 25, 140–142
    convergence and stability of, 154–158
    for discontinuities, 127
    with explicit R-K two-stage process, 59–60
    polynomial, 148–149
    rational, 150–154
    for singularities, 127, 136–137
    for stiff problems, 164, 198, 200, 217–223
    zero-th column for tables for, 143–148
    See also Richardson extrapolation

F

FACSIMILE code, 213
Fehlberg pair approach
    coefficients for, 67
    for R-K local error estimates, 54–55

Finite difference equations, 10–13
Finite differential methods, 245
First integral mean value theorem, 17
Fixed coefficient formulas
    with Adams methods, 123
    for variable step LMMs, 214–215
Fixed leading coefficient formulas, 214
FORSIM code for discontinuities, 26, 242
Forward difference operator, 2

G

GEAR code, 119
    for singularities and discontinuities,
        127, 136, 139
    for stiff/nonstiff differential systems,
        213–214
GEARIB code, 213–214
Gill four-stage Runge-Kutta process,
    matrix for, 66
Global errors, 28–30
    estimation techniques for, 244–245
    with Euler schemes, 40
    with LMMs, 103, 109, 112
    of one-step schemes, 33
    with R-K schemes, 54
Global extrapolation
    with implicit midpoint rule, 146
    with trapezoidal rule, 144
Gragg-Bulirsch-Stoer extrapolation
    scheme, 151–153
    with singularities, 136
    for stiff systems, 217
GRK4A code
    for Rosenbrock method, 87
    for stiff algorithms, 206
GRK4T code for stiff algorithms, 206

H

Hammer and Hollingsworth implicit
    Runge-Kutta scheme, 77
Heun Runge-Kutta processes, 58
    three-stage matrix for, 64
    two-stage matrix for, 63
Hybrid methods for numerical
    integration, 25

I

IMPEX code for extrapolation tables, 146
IMPEX2 code for stiff algorithms, 200
IMPEX3 code for stiff algorithms, 200
Implicit Euler schemes
    with discontinuities, 139
    with LMMs, 99
    for stiff IVPs, 163
Implicit linear multistep methods, 90,
    95–100
Implicit midpoint rule
    for extrapolation tables, 142, 146–147
    and IMPTR-3, 46
    for stiff systems, 217–218
    for time-dependent problems, 181
Implicit Runge-Kutta schemes, 77–79, 85,
    239
    coefficients for, 50, 224
    computation steps needed with, 181
    and Dahlquist Barrier Theorem, 164
    stability with, 173, 188–195
    with stiff algorithms, 200, 203–211
    two-stage, 75–76
IMPTR-1 code, trapezoidal schemes
    using, 43–44
IMPTR-2 code, A-stability of, 45
IMPTR-3 code and implicit midpoint
    rule, 46
IMSL code and R-K local error estimates,
    54
Increment function of one-step schemes,
    60–61
Influence function approach with LMMs,
    95, 111–112
Interpolation theory, 3–10
INTRP code for discontinuities, 27
Inverse Euler scheme, 37
    for extrapolation tables, 142–143, 146
    with rational extrapolation, 151–152
    stability of, 157–158
Inverse polynomial methods with
    singularities, 133–136
IRK. *See* Implicit Runge-Kutta schemes
Iteration schemes
    with LMMs, 101
    for numerical integration, 21–22
    order of, 168–169
    with stiff problems, 169–171, 211–212,
        224

J

Jacobian estimate with nonlinear IVPs, 170–172

K

Kronecker delta, 4
Kutta Runge-Kutta processes, matrix for, 65
Kutta-Merson Runge-Kutta schemes, 74
matrix for, 66

L

Lagrange interpolation scheme, 3–7, 9–10
Least squares multistep formulas, 200
Lie series approach, 21–22
Linear algebra, reduction of, 239–240
with Rosenbrock method, 85
Linear Multistep methods, 25
Adams codes for, 119–124
A-stability in, 164, 182
explicit, 91–95
general theory of, 104–118
implicit, 95–100
partitioning with, 241
predictor-corrector methods with, 100–104, 173–174
RAS for, 173, 175, 180
for second order differential equations, 228–230
starting procedure for, 89–91
Linear systems with constant coefficients, 13–15
Lipschitz condition, IVP satisfaction of, 24
LMM. *See* Linear Multistep methods
Local errors, 28–30
with Adams-Moulton scheme, 122
in discontinuous systems, 127, 136, 139
with extrapolation processes, 158
with LMM predictors, 103
for R-K methods, 54–55, 74
for singularities, 127
Local extrapolation
with Adams-Moulton scheme, 122–123
with rational extrapolation, 154

with R-K methods, 55, 73–74
with Rosenbrock method, 88
with trapezoidal rules, 144
Local truncation errors
with Adams-Bashford scheme, 94
with Adams-Moulton scheme, 96
with discontinuities, 139
with LMMs, 104, 109, 112
with nonstiff algorithms, 167
of one-step schemes, 31–32
with R-K methods, 51–53, 73
for singularities, 128, 135
with Taylor expression for IVP, 48
Logarithmic norm and stability of
one-step methods, 35
Lotkin error bound, 52
LSODA code, 120, 202
for stiff/nonstiff differential systems, 213–214
LSODAR code, 202
for discontinuities, 126, 242
root finder in, 26
for stiff problems, 213–214
LSODE code, 120, 238
for stiff problems, 27, 202, 208–210, 213–215
LSODES code, 171, 202, 213–214
L-stability
of numerical integrators, 182
with Rosenbrock method, 86
with semi-implicit Runge-Kutta scheme, 79, 81–83
with stiff problems, 223–224

M

Matrix representations
of R-K processes, 56–60, 62–71, 76
of scalar iterations, 88, 212, 215
Maximum norm, 18–19
Mean Magnitude norm (MM), 18–19
Mean Value Theorem, 17
with LMMs, 112
MEBDF code
angle of stability of, 212
for stiff problems, 27, 211, 213
Merson five-stage scheme, stability with, 172–173
Meshpoints
and points of discontinuity, 26

of trapezoidal rule for extrapolation
tables, 144
Meshsize
with Adams-Moulton scheme, 123
and convergence, 30
with MIRK, 224
for R-K explicit two-stage process, 59
and stepsize selection, 242
with stiff problems, 27, 202
METAN1 code
for semi-implicit midpoint rule, 222,
239
for stiff problems, 27, 200
METAN2 code for semi-implicit midpoint
rule, 223, 239
Midpoint rule for Runge-Kutta method,
matrix for, 63. *See also* Explicit
midpoint rule; Implicit midpoint rule
Milne error estimates with LMM
predictors, 103, 105–108
Milne procedures for implicit LMMs, 97
MIRK code
angle of absolute stability of, 185
with semi-implicit R-K schemes, 84
for stiff problems, 223–224
MM (mean magnitude) mode, 18–19
Modified midpoint rule
with extrapolation, 158
for extrapolation tables, 143, 145–146
Mono-implicit Runge-Kutta scheme, 81,
239
coefficients for, 224
for stiff problems, 200, 223–225
Multiderivative methods
multistep, 164, 174, 178–180
for second order IVPs, 234–237
with singularities, 129
Multiple shooting methods, 245

N

Neighboring problem, 244
Newton backward interpolation formula,
8–9
Newton bases, 9
Newton divided difference scheme, 5–9
Newton scheme for stiff IVPs, 169–171
Newton-Raphson scheme
with stiff problems, 216
with trapezoidal schemes, 44

Nonlinear equations, solution of, 168–171
Nonlinear Multistep methods, 25
with singularities, 135
for stiff IVPs, 164
Non-polynomial methods for singularities,
127–133
Nonstiff algorithms, 167–168
Nordsieck array with Adams methods,
124
Norms, common, 18–19
Numerical differentiation approach to
LMMs, 98–100
Numerical integration algorithms, 20–22
A-stability of, 181–184
characteristics of, 28–30
existence of solutions with, 23–25
with special IVPs, 26–28
for stiff IVPs, 163
Nystrom methods
explicit LMM, 97–98
periodicity of, 230
RAS for, 177
R-K schemes, matrix for, 65

O

ODIOUS code for stiff problems, 211
angle of absolute stability of, 185
One-leg hybrid methods, 232–236
One-leg multistep methods, 181, 195–197
for second order problems, 232–236
One-norm, 18–19
One point iteration scheme, 168–169
One-step schemes, 31–34. *See also* Euler
schemes; Richardson extrapolation;
Trapezoidal schemes
Optimal Runge-Kutta processes
three-stage, matrix for, 64
two-stage, 58, 63
Optimal zero-stable LMMs, 115
Order
of A-stable LMM method, 45
of integration formulas, 33–34
of LMMs, and Dahlquist Barrier
Theorem, 115
with P-stable methods, 231
for R-K processes, 51
for Rosenbrock methods, 86
of second derivative formulas, 216
of semi-implicit R-K methods, 205
with Taylor expression method, 49

Oscillatory problems, 28
  cyclic methods for, 242–243
  solution of, 168–169
  with stiffness problems, 166, 206,
    209–211

P

Pade approximation with stiff IVPs, 34
Partitioning and linear algebra, 240–241
PASVA3 code, 245
PEC mode
  with LMMs, 100, 105–106
  stability of, 174
PECE mode
  with LMMs, 101, 106–108
  Simpson rule in, 116
Periodicity
  of LMMs, 229–230
  for multiderivative methods, 234–237
Perturbed polynomial techniques,
    126–127
Pi, extrapolation process to determine,
    140–142
Picard iteration scheme, 21–22
Polynomial extrapolation method,
    148–149
  stability of, 156–158, 193
Polynomials, root distribution for, 15–17
Predictor/corrector methods
  absolute stability with, 173–174
  with LMMs, 100–104
  with singularities, 131–133
PSA (pseudostationary approximation),
    240
P-stability, 230–231
  formulas for, 232–236

R

R-K. *See* Runge-Kutta schemes
Ralston scheme, matrix for, 64
RAS. *See* Region of absolute stability
Rate of convergence for iteration schemes,
    171
Rational extrapolation method, 150–154
  stability polynomial for, 193
Rational function methods, 126
Recurrence relation of trapezoidal
  schemes, 46

Region of absolute stability, 172–180
  with Adams-Moulton scheme, 123
  with nonstiff algorithms, 167
Relative stability of LMMs, 118
Remainder Theorem with LMMs,
    109–111
Restarting of integration with
  discontinuities, 26, 139
RFK45 code
  for nonstiff methods, 167–168
  and R-K local error estimates, 54–55
  for stiff algorithms, 202, 205
Richardson extrapolation
  for Euler schemes, 37–39
  with R-K methods, 59–60 74, 203
RICM4 code, 227
RICM6 code, 227
ROC (rate of convergence) for iterative
  schemes, 171
Root Mean Square (RMS) norm, 18–19
Rooted trees for R-K processes, 51
Roots of polynomials, distribution of,
    15–17
Rosenbrock methods, 85–88, 239
  semi-implicit R-K schemes, 79
  for stiff algorithms, 200, 205
Rosenbrock Wanner methods, 86–88
Roundoff errors, 30
ROW (Rosenbrock Wanner) methods,
    86–88
Runge-Kutta schemes, 25, 48–52
  convergence and stability of, 60–62
  discontinuity starter with, 139
  error estimation for, 52–55, 72–73
  explicit midpoint rule as, 149
  explicit two-stage process, 56–60
  four-stage, matrices for, 65–66
  implicit and semi-implicit, 75–85
  for LMM starting values, 90
  matrix representations of, 62–71
  meshsize for, with stiff algorithms, 202
  one-step, matrix for, 63
  partitioning with, 241
  Rosenbrock methods, 85–88
  second order, 236
  seven-stage, matrix for, 68–70
  six-stage explicit, for LMM starting
    values, 90
  and stepwise selection, 73–75
  for stiff algorithms, 198, 200
  three-stage, matrices for, 64–65

two-stage, matrices for, 63–64
See also Explicit Runge-Kutta schemes;
  Implicit Runge-Kutta schemes;
  Mono-implicit Runge-Kutta scheme;
  Semi-implicit Runge-Kutta schemes
Runge-Kutta-Nystrom scheme, 227, 230

S

Scaled iteration matrices, 88, 212, 215
Schur criterion, 15
Schur polynomials, 16–17
  for RAS for LMMs, 180
SDBASIC code with second derivative
  formulas, 216
SDF. *See* Second derivative formulas
SDIRK formula
  angle of absolute stability of, 185
  and IVP stiffness, 27, 208
  with semi-implicit R-K schemes, 80
SDRROOT code, angle of absolute
  stability of, 185
Second derivative formulas
  partitioning with, 241
  stability polynomial for, 193
  for stiff problems, 27, 198, 200,
    215–217
Second order differential equations, 57
  LMMs with, 228–230
  multiderivative methods with, 236–237
  one-leg methods with, 232–234
  and P-stability, 230–231
Semi-implicit midpoint rule, 58, 239
  for extrapolation tables, 143, 147–148
  for stiff systems, 217–223
Semi-implicit Runge-Kutta schemes,
    77–84, 239
  coefficients for, 50, 224
  computation steps needed with, 181
  stability polynomial for, 193
  for stiff algorithms, 200, 205–206
  three-stage, 79, 81
  two-stage, 78, 79, 81
Shift operator, 1–2
SIMPLE code, 239
  for stiff problems, 27, 211
Simpson rule
  error constant for, 111–112
  as optimal LMM, 115–116
Singly implicit Runge-Kutta method, 200,
    206–208, 239

Singularities, 126
  and extrapolation, 140–159
  inverse polynomial for, 133–136
  non-polynomial methods for, 127–133
SIRK (singly implicit Runge-Kutta)
  method, 200, 206–208, 239
Smoothing procedure with modified
  midpoint rule, 145
Solutions, existence and uniqueness of,
    23–25
S-stability of one-step schemes, 182
Stability, 29–30
  of Adams-Bashford method, 95
  of Adams-Moulton scheme, 123
  algebraic, 188, 192–195
  of cyclic methods, 243
  with distribution of polynomial roots,
    15–17
  of Euler schemes, 36, 157
  of explicit R-K two-stage scheme, 82
  of extrapolation processes, 156–158
  of fixed coefficient formulas, 215
  of implicit R-K schemes, 77, 85,
    188–195
  of inverse Euler scheme, 37, 157
  with iteration methods for LMMs, 101
  of LMMs, 104, 115–118, 229
  of L-stable R-K schemes, 84
  of multiderivative methods, 179, 236
  with nonlinear systems, 171
  of nonstiff algorithms, 167
  with one-leg multistep methods, 196
  of one-step schemes, 33–35, 182
  with PEC mode, 101
  of polynomial extrapolation, 149
  of R-K methods, 55, 61–62, 188–192,
    236
  of Rosenbrock method, 86, 205–206
  of second derivative formulas, 216
  with semi-implicit midpoint rule,
    219–221
  of semi-implicit R-K schemes, 78
  for SIRK method, 207–208
  and Stetter's prediction, 116
  and stiffness, 166, 180–188
  with trapezoidal schemes, 43–45
  See also Absolute stability; Angle of
    stability; A-stability
STEP code, 120
  for discontinuities, 126, 242
  for nonstiff IVPs, 167

Stepsize selection
  with DIFSUB, 26
  for implicit R-K schemes, 203
  and meshsize, 242
  with rational extrapolation, 154
  for R-K schemes, 73–75
  with stiff algorithms, 201
Stetter predictor, 115–116
Stiff algorithms, 167–168, 198–203
  BDF, 211–215
  extrapolation processes, 217–223
  implicit R-K schemes, 203–211
  mono-implicit R-K schemes, 223–225
  second derivative formulas, 215–217
Stiff initial value problems
  absolute stability with, 172–180
  algorithms for. *See* Stiff algorithms
  and concept of stiffness, 161–167
  and Euler scheme, 37
  and explicit R-K processes, 75
  and implicit R-K processes, 188–195
  and meshsize, 27
  nonlinear equations for, 168–172
  one-leg multistep methods for, 195–197
  oscillatory, 166, 206, 208
  stability criteria for, 180–195
  Taylor series expansion for, 22
Stiff/nonstiff differential systems, codes
    for, 213–214
Stiffness ratio, 165
STINT code
  angle of absolute stability of, 185
  for stiff problems, 27, 208, 210
Stormer-Cowell method, 228
STRIDE code, 239
  angle of absolute stability of, 185
  for SIRK, 208
  for stiff problems, 27, 208–210
Strong stability of LMMs, 118
Substitution methods with second order
    problems, 233–234
Switching techniques, 242
  for discontinuities, 26–27, 126–127
  for singularities, 126–127
Symmetric multistep methods, 230

T

Taylor series method
  for discontinuities, 27

with explicit R-K two-stage process,
    56–57
  for IVPs, 48–49
  for L-stable R-K schemes, 83
  for numerical integration, 21–22
Transient phase, 28
  with nonstiff algorithms, 167
  for scalar IVPs, 161–162
TRAPEX code for stiff algorithms, 200
Trapezoidal schemes, 42–47
  with discontinuities, 139
  for extrapolation tables, 142–144
  implicit R-K schemes, 76
  matrix for, 76
  stability with, 43–45, 181
  for stiff systems, 217
Trees
  for order relations for Rosenbrock
      methods, 86
  rooted, with R-K schemes, 51

U

Unconditional stability, 230
Uniqueness theorem, 24–25

V

Van der Pol oscillator, 21
Variable coefficient approach
  with Adams methods, 123
  for variable step LMMs, 214–215
Vector form of IVPs, 21
VOAS code, 119

W

Weak stability of LMMs, 115, 118

X

XRK code
  coefficients for, 70–71
  for R-K methods, 55

Z

Zero-stability of LMMs, 115–118
Zero-th column of extrapolation tables
  generation of, 143–148
  semi-implicit midpoint rule for, 220